A History of Land Use in Mongolia

A History of Land Use in Mongolia

The Thirteenth Century to the Present

Elizabeth Endicott

palgrave
macmillan

A HISTORY OF LAND USE IN MONGOLIA

First published in 2012 by
PALGRAVE MACMILLAN®
in the United States—a division of St. Martin's Press LLC,
175 Fifth Avenue, New York, NY 10010.

Where this book is distributed in the UK, Europe and the rest of the world, this is by Palgrave Macmillan, a division of Macmillan Publishers Limited, registered in England, company number 785998, of Houndmills, Basingstoke, Hampshire RG21 6XS.

Palgrave Macmillan is the global academic imprint of the above companies and has companies and representatives throughout the world.

ISBN: 978–1–137–26965–2

Library of Congress Cataloging-in-Publication Data.

Endicott, Elizabeth.
A history of land use in Mongolia : the thirteenth century to the present / Elizabeth Endicott.
p. cm.
Includes bibliographical references and index.
ISBN 978–1–137–26965–2 (alk. paper)
1. Land use—Mongolia—History. 2. Land use, Rural—Mongolia—History. 3. Rangelands—Mongolia—History. 4. Herders—Mongolia—History. I. Title.

HD920.8.E53 2012
333.73′1309517—dc23 2012016618

A catalogue record of the book is available from the British Library.

Design by Newgen Imaging Systems (P) Ltd., Chennai, India.

First edition: November 2012

10 9 8 7 6 5 4 3 2 1

Transferred to Digital Printing in 2013

Contents

Illustrations

All photographs by Elizabeth Endicott

Acknowledgments

First and foremost, I am deeply indebted to the Mongolian interpreters, local guides, and driver-mechanics upon whose skills I relied during my 1996, 2002, 2005, 2006, 2010, and 2011 visits to the Mongolian countryside. Without their field support and interest in my research, I would never have been able to travel as widely in rural Mongolia. For travel to Mongolia in 2002 and 2006, I received Middlebury College Faculty Professional Development Grants for which I am most grateful. Over the years, I have been most fortunate to have had the financial support of my family, and, on several occasions, I have enjoyed the company of family members on my journeys into the Mongolian countryside.

I would also like to acknowledge the following individuals for taking the time to exchange ideas, opinions, and information on Mongolia: Robin Charpentier, former director of the American Center for Mongolian Studies (ACMS) in Ulaanbaatar; the late O. Pürev who in 2005 took a day to talk about Mongolia's historical geography with me; B. Erdenebolor and B. Nyamgerel of the Mongolian State University of Agriculture; the staff of the Land Office in Ulaangom; Ch. Davaadalai; Annika Ericksen; Sas Carey; Anne Cordes; Professor Emerita Esther Jacobson-Tepfer; and Evan West and Ethan West for their enthusiasm for my research project. Ethan West offered crucial technical support in formatting the manuscript along the way. Michael A. Paretti's thoughtful suggestions for selecting photographs from my voluminous personal archive of Mongolia photos from the years 1978 through 2011 were very helpful. I am grateful to Ben Meader of the Middlebury College Geography Department for producing the two maps in this book. A special acknowledgment goes to Professor Emeritus Thomas T. Allsen who read the entire manuscript and offered incisive comments resulting in important revisions. Responsibility for any errors in the book, however, remains entirely with the author.

During the research process, I benefited from comments on my presentations on land use in Mongolia, specifically at the ACMS Speaker's Series in Ulaanbaatar in 2005; the International Congress of Mongolists in Ulaanbaatar in 2006; the Faculty Lecture Series at Middlebury College in 2007; and the International Congress of Mongolists in Ulaanbaatar in 2011.

Finally, a disclaimer: this book does not pretend to offer generalizations about all of Mongolia. For instance, I have never visited Bayan-Ölgii Aimag in Mongolia's far west nor have I visited Dornogov' or other areas of the eastern Gobi. At a certain point, however, one simply has to write what one knows and let others pick up the storyline in the future…

ABBREVIATIONS

EMME	Christopher P. Atwood. *Encyclopedia of Mongolia and the Mongol Empire* (New York: Facts on File, 2004).
MM	*The Mongol Messenger* (weekly English-language print newspaper published in Ulaanbaatar, Mongolia).
MTW	*mongoliathisweek* (now defunct online English-language newspaper published in Ulaanbaatar, Mongolia), http://www.mongoliathisweek.mn/lanl.htm.
MWN	*Mongolia Web News* (English-language news online), http://www.mongolia-webnews.mn.
SDC	Swiss Agency for Development and Cooperation in Mongolia, http://www.swiss-cooperation.admin.ch/mongolia/.
UBP	*UB Post* (online Ulaanbaatar newspaper), http://ubpost.mongolnews.mn.

CHAPTER 1

INTRODUCTION

WHY WRITE A HISTORY OF LAND USE IN MONGOLIA? Certainly it would be worth correcting the popular misconception of nomads "roaming" or "wandering" across the steppes of Inner Asia, a notion that still prevails as I found every spring while teaching my "Nomads of Eurasia" course at Middlebury College. Any nomadic group that moves its herds without carefully planned migratory routes with sufficient water and pastureland along the way would be doomed. In spite of the much repeated stereotype in Chinese and other sedentary historical sources that nomads "wandered in search of water and grass," the truth is that pastoral nomadic societies in Central Eurasian history have been very well organized to survive—and at times thrive—in the inhospitable deserts and grasslands of this part of the world. At the very least, this book may dispel for once and all the notion that nomads aimlessly "wander" in a sort of timeless haze impenetrable to historical analysis.

This book offers a historian's perspective on several key questions: how Mongolia's nomads have used the natural resources of their homeland over the past eight centuries or so; to what degree outside forces have shaped herders' socioeconomic systems; and how the recent (post-1990) painful transition from herding under socialist state supervision to a relatively unregulated way of life in the era of a market-driven, profit-oriented economy has impacted the pastoral nomadic experience in the twenty-first century. These questions are complex, and there are no easily formulated solutions to ensure the future survival of Mongolia's pastoral nomadic herders in a nation that is experiencing dramatic economic growth through the development of its mineral wealth.

As the world's least densely populated country, with about 2.8 million people spread across 1.5 million square kilometers, Mongolia is also notable for the size of its national livestock herd. At the end of 2010, Mongolia was home to some 31.8 million head of domestic livestock that fall into the categories of what the Mongols call *tabun khoshighun mal* (five types of herd animals): horse, camel, cow (including yak), sheep, and goat. Picture a country slightly smaller than Alaska or almost half the size of India with over a third of its 2.8 million population living in the capital city of Ulaanbaatar and up to 40 percent of its population making a living from mobile herding (often referred to as pastoral nomadic herding).

The Mongolian countryside is strikingly devoid of fencing, and the land is dotted with only the smallest of structures for storage and for winter/spring livestock shelters. Because hay and other feed are only supplementary for livestock, the land itself is the key resource in herding. But how is access to pastureland determined? This book will attempt to grapple with the question of how rights to land use in these vast open spaces have been determined in the past and in the present.

Mongolia's natural beauty is hard to ignore in a discussion of land use, and thus this book incorporates a few of the photographs taken by the author over the course of the past 33 years. Much has changed between my first trip to Mongolia in 1978 and my most recent visit in 2011. Change and continuity are interwoven in the images. With the intent of drawing the reader into the landscape, the photographs also will show key elements of day-to-day nomadic life that dictate many aspects of land use among Mongolia's pastoral nomadic households. All the photographs in the book were taken by the author.

Because a substantial portion of this book is devoted to the post-1990 circumstances of Mongolia's herders, the findings should be of interest to the Mongolian people themselves. While it is doubtful that the herders whose historical experience is the centerpiece of this work will have easy access to the book, it is not improbable that at least a few Mongolian policymakers may find it an interesting read. Thus, it is to the people of Mongolia that I dedicate the present book. I have enjoyed the legendary hospitality of the Mongolian people over the past 30 some years while traveling in their grasslands, mountains, and desert, and I offer this book by way of thanks.

Theoretical and Practical Questions

Mongolia's 2006 annual livestock census raised alarm among some observers concerning the twin problems of overgrazing and pasture deterioration. According to this census, Mongolia's national livestock herd had reached 34.48 million head, at that point the highest number since 1924.[1] The 2006 census raised important questions: were Mongolia's nomads good stewards of the land? Who should oversee animal-to-pasture ratios? Was Mongolia witnessing unprecedented issues in land use, particularly the immense growth of goat herds in response to the international market in cashmere? Do *zuds*, the cruel cyclical climatic disasters that cause severe loss of herds, actually serve to correct the imbalances in herd-to-pasture usage that may arise?

Interestingly, the government of Mongolia applauded the announcement of herd growth as a positive development. At a New Year's Eve gala event at the foot of the mammoth statue of Chinggis Khan (completed in late 2006) in Sukhbaatar Square in the nation's capital, Ulaanbaatar, Mongolia's president N. Enkhbayar congratulated the nation on "having achieved much in the socio-economic sector." His speech included this comment on national herd size:

> The national herd now numbers over 34 million and vegetable production has increased by up to 60 percent. GDP has risen about 8 percent and the national budget showed a surplus of over Tg300 billion [US$1 = 1,298 Mongolian Tögrög]. The future of the economy looks brighter.[2]

The larger the herd, the better the nation's economic prospects? It may be difficult to break this habit of thought both among government officials and among herders themselves. The Mongolian government's bestowing of awards upon herders with more than one thousand head of livestock and issuing of cash payments for goats after the 2008 cashmere price crash reinforced two dubious trends: the overall increase in livestock and the disproportionate increase in goats, the animal most responsible for rangeland deterioration.[3] The 2004 Open Society Forum's survey of herder opinion reflected the same equation of numbers of animals with improved circumstances. Increasing the number of livestock was the first priority among 773 surveyed herders, with improving the breed of livestock as a close second.[4] In her beautifully written and photographed 2010 volume on Mongolian nomads, the French sociologist Linda Gardelle recorded numerous conversations with her herder informants. These conversations universally attested to the attitude that increasing herd numbers was vital to one's well-being: "Livestock are our bank account, our money"; "To improve our lives, we increase our livestock"; and "Thanks to livestock, we live well. Many livestock means security."[5] Clearly, herd size has been and continues to be an indicator of success as defined by herders themselves. The Mongolian government contributes to this equation also.

During my summer 2011 trip through Uvs Aimag in northwestern Mongolia, herders would point out with pride a *miangat malchin* (a herder household with one thousand or more head of livestock). A miangat malchin earns a government certificate, medal, and cash for this accomplishment. This relic of the socialist era in Mongolian history—the awarding of medals and material rewards for large herds—encourages large herd size, particularly of goats.

Mongolia's herders have historically been motivated to maximize livestock numbers in good years in order to offset losses in bad years. This strategy is understandable, given the periodic recurrence of zud, the Mongolian term that encompasses disasters caused by unusually heavy snows that prevent livestock from accessing forage, icing over of pastures, prolonged cold—all winter and spring phenomena—or summer drought. Recent field studies that have included in-depth interviews with herders in different *aimags* (provinces) demonstrate that the mentality of increasing the size of herds is deeply ingrained and likely to continue.[6] Yet, uncontrolled growth of livestock herds, particularly the increase in goats that destroy grass at its roots while grazing, certainly results in pasture degradation, which, in turn, will contribute to the impoverishment of the herders themselves.

In terms of sheep and goats, it is generally accepted that the ideal herd composition in Mongolia would be three sheep to one goat. In the year 2000, however, the ratio of sheep to goat stood at 1.2:1; in 2008, goats outnumbered sheep. Goat numbers hit a high of 19,600,000 in 2009.[7] Cashmere, the fine wool from goats' undercoat, provides herders with significant income, and the international market for this luxury item explains the spike in goat numbers. By the end of 2010, however, zud conditions led to a great reduction in goat numbers, reversing, for the first time since 2005, the trend of goats outnumbering sheep. Zud and livestock epidemics were held responsible for the loss of more than ten million herd animals in 2010; the national herd was, by the end of 2010, 27.7 percent less than it had been in 2009.[8] As is typical in the recurring cycle of losses and gains in a pastoral nomadic economy, the preliminary 2011 statistics showed a significant rebound in livestock numbers. Mongolia's national herd at the end of 2011 had increased by 11 percent over 2010. Once again, the largest increase was in the number of goats, with a 14.8 percent increase over 2010.[9]

When we step back for perspective and look at the years prior to the 2009–2010 winter zud, we see the clear pattern of steady growth in livestock population. The year 2007 saw an 18 percent increase in the national herd; Mongolia's rangelands by the end of that year were supporting 40.3 million herd animals.[10] Over a six-year period, from 2002 to 2009, Mongolia's national herd increased by an astounding 85 percent.[11] In 2009, the livestock number grew to 44 million head.[12] As a recent analysis of this problem stated, "The herders' interest in maximizing livestock in the current incentive structure is a primary challenge to building sustainable rangeland management."[13] Can this trend continue unabated?

The optimal size of herd is, of course, closely tied to the carrying capacity of the land itself. Carrying capacity refers to "the stocking rate that achieves a targeted level of animal or economic performance over a defined period of time without causing a deterioration of the pasture ecosystem."[14] Dire warnings about saturation of Mongolia's grazing lands have periodically been raised over the past century. As early as 1929, the eminent Russian geographer, A. D. Simukov, wrote of a "pastureland crisis" that was close at hand.[15] Simukov mused that "it was entirely possible to envision Mongolia as a country saturated with livestock, and in the very near future extensive livestock-raising will be at a dead end."[16] In spite of Simukov's well-deserved reputation as an astute observer whose fieldwork findings continue to provide invaluable data and a "feel" for many issues of pastureland usage, his 1929 prediction was not borne out in the twentieth century.

Simukov's ringing of the alarm over pastureland saturation has been repeated by a number of other observers in more recent decades. For instance, a 2009 study, under the auspices of the World Bank's Netherlands-Mongolia Trust Fund for Environmental Reform (NEMO), laments pasture deterioration owing to the growing livestock numbers (particularly goat herds) and a drier climate.[17] Certainly the imbalance in herd structure caused by an overabundance of goats has been a major contributing factor to pastureland deterioration. It is entirely understandable that herders were only following global market trends that favored cashmere production in the 1990s; fluctuations in cashmere prices may in the long run induce herders to restructure and balance their herds.

Predictions of imminent crisis may reflect not only one's perception of actual conditions, but also one's preexisting suppositions about pastoral nomadism's ultimate chances of survival. As one scholar notes, "Ecology-framed 'end-of-pastoralism' scenarios have dominated the history of pastoral development."[18] The so-called end-of-pastoralism scenarios generally hinge on resource scarcity (i.e., pasture depletion, water shortage, etc.), global climate change, or violent conflict (the latter applicable to certain pastoral nomadic groups in Africa). In Mongolia's case, predictions of the gradual petering out of the pastoral nomadic way of life tend to combine resource scarcity with the well-documented effects of climate change.

Pastureland saturation by livestock is just one area of concern in the larger issue of the long-term viability of this way of life. For instance, whenever a zud occurs, in-country observers as well as foreign observers tend to predict the demise of the pastoral nomadic way of life. The zud that hit the western aimags of Mongolia with heavy snowfall in January 2009 led a *UB Post* reporter to conclude that "such natural disasters could spell catastrophe for the herder's way of life."[19] A World Bank representative who, in early February 2010, toured areas of Mongolia that had been hard hit by the most recent and worst zud in decades opined that environmental degradation caused by current herding practices, including overstocking, contributed as much as the weather did to the enormous losses of livestock in winter/spring 2010: "Can fragile ecosystems like those in Mongolia continue to bear the burden of an ever increasing livestock herd that continues to deplete pastures and threaten long run sustainability?"[20] As noted earlier, the 27.7 percent decrease in national herd between 2009 and 2010 (from 44 million head to 31.8 million head) was largely the result of zud.

The ferocity of the 2010 zud that affected over 80 percent of Mongolia's territory predictably fed the debate over pastoral nomadism's chances of survival in the twenty-first century. As Troy Sternberg's article on the 2010 zud phrased it, "The *dzud* [zud] will cause a re-evaluation of the role and viability of pastoralism which frames the country's cultural identity and social traditions."[21] Sternberg further predicted that the post-zud flight of herders who suffered catastrophic livestock losses to Ulaanbaatar and smaller urban centers would lead to a downsized herding sector; herding would then evolve into "an idealized livelihood rather than a dominant profession."[22]

This scenario assumes a one-way, irreversible rural-urban migration, an assumption that has not always proven true. When hard economic times hit urban centers, a reverse flow from urban to rural may occur, as was well documented in the decade of the 1990s when the number of registered herders tripled, surging from 135,420 in 1989 to 407,030 in 2001. By 2001, herders constituted almost 50 percent of Mongolia's working population.[23] More recent in-depth fieldwork has identified the same urban to rural flow of households in Ikhtamir *Sum* (county) in Arkhangai Aimag where the number of herder households increased from 316 in 1990 to 1,100 in 2008.[24] Economic pressures, particularly high unemployment rates in the sum center and other urban centers motivated people to turn to or return to herding as a livelihood. Herding thus has served as a crucial economic safety net after the USSR-subsidized economy of socialist Mongolia collapsed. A return to herding, while not successful for all new herders, has allowed many Mongols to escape from rampant urban unemployment, food shortages, and the sudden end of state-funded social services in the 1990s.

While zud have contributed to the debates over pastoral nomadism's viability in modern times, man-made ideologically driven schemes have also threatened this way of life. In the socialist era in Mongolia (1920 to 1921), the eventual phasing out of pastoral nomadic production was trumpeted as a worthy goal; industrialization and urbanization were imagined as steps to Mongolia's modernization in the twentieth century. Pastoral nomadism was described as "archaic" and the socialist mission was "to transform the MPR in the near future into an industrial-agrarian country" as one Soviet-era Mongolist wrote in 1979.[25] V. V. Graivoronskii saw the end of the nomadic way of life on the historical horizon: "The settling of nomads is a result of a gradual progressive development of human society from more simple and primitive forms of production and existence to more complex and progressive [forms]."[26] A sense of inevitability pervaded the writings of Graivoronskii and other Soviet scholars in their predictions of the demise of the pastoral nomadic way of life.

Yet, the rapidly changing social reality in the two decades since the end of the socialist era has led some naysayers into a reconsideration of the viability of pastoral nomadism. Graivoronskii, for one, has transformed his approach to pastoral nomadism's past, present, and future in Mongolia. Writing in 2007, he ruminated that because of pastoral nomadism's complexity, the solution to its problems will not be found in short-term approaches, such as forcing nomads into a settled way of life. Looking back over the twentieth-century attempts to do just this in the former USSR and in Mongolia, Graivoronskii admits to the resulting damage to livestock production.[27] Sadly, these historical lessons continue to be ignored in other pastoral zones of the world: in ethnically Tibetan regions of China, for instance, news reports continually document the forced settlement of livestock herders. Chapter 5 will explore the ramifications of Chinese state policy toward pastoralists.

Other researchers view the developments of the postsocialist era (1990 to the present) as contributing directly to pastoral nomadism's likely demise. These researchers highlight the fragility of rangeland management institutions after the collapse of a centralized socialist management system, and raise questions about the viability of pastoral nomadism. For instance, the Danish anthropologist Ole Bruun wrote in 2006 that "pasture management and facilities for the dispersion of grazing pressure are in ruins."[28] Bruun's observations are very valuable, and his conclusions about the weak governmental role in pasture resource management are complemented by Troy Sternberg's critique of weak governmental response to zud disasters and the resulting social fallout.[29] As the second decade of the twenty-first century progresses, there are clear indications that the role of government in managing pastureland may grow. Chapter 7 will delve into the pending legislation on pastureland that is under discussion in the Mongolian parliament.

Debates over the Mongolian "livestock crisis" continue. A 2005 study, for instance, discusses "the Mongolian livestock crisis," while detailing what a slippery concept "carrying capacity" actually is, owing to the "crude and static" rules of measurement that are often employed.[30] Many specialists on the issue of Mongolian pastureland's carrying capacity do believe that the maximum has already been reached and exceeded. Enkh-Amgalan Tseelei, a Mongolian researcher for the Swiss Agency for Development and Cooperation, estimates that Mongolian pastureland's carrying capacity has been exceeded by 60 percent.[31] Overgrazing of course is enormously damaging to pastureland. If a recent estimate that 70–80 percent of Mongolia's grazing lands have already been degraded is accurate, then the debates over land use and strategies for pasture management certainly acquire even greater urgency.

The startling array of statistics all documenting increased numbers of livestock—for instance, from 23,771,400 in 1980 to 43,228,400 in 2008—should cause great concern over the sustainability of the pastoral nomadic way of life in Mongolia.[32] Yet, climate change is also a factor in pasture degradation, since recent evaluations have shown that the trend toward pasture deterioration began decades before the post-socialist era's explosive growth in herd numbers.[33] In fact, a distinct decrease in annual precipitation over the past 65 years in Mongolia has left its mark on the fragile ecosystems of the country's pasturelands. According to Linda Gardelle's extensive interviews, both Mongolian herders and foreign experts expressed more concern about the effects of climate change than about any of the other numerous problems in the realm of pasture and water resources.[34]

Thus, the threat posed by climate change looms larger than political or economic meddling from authorities outside the herders' domain. Since the demise of the socialist experiment in Mongolia, its political leaders have rarely called for sedentarization of Mongolia's herding population. Two decades of working in Mongolia may have finally tempered certain foreign aid organizations' initial naïve enthusiasm for privatizing pastureland and converting nomadic herding to Western-style ranching.

While Mongolia's pastoral nomadic population undeniably has faced crises through the ages, the particular challenges of the postsocialist market era include, but are not limited to, the carrying capacity of pastureland. Over the centuries, Mongolia's pastoral nomads have been influenced to varying degrees by different forms of authority: clan/tribal; imperial; religious; and, more recently, secular. Mongolian nomads' historical relationships with authority imposed from the outside have parallels with nomad-state relationships in other parts of the world. The well-developed literature on Israeli-Bedouin interactions is a case in point. In an article entitled "Are There Pastoral Nomads in the Arab Middle East?," the Israeli anthropologist Emanuel Marx explored the reasons why governments in various parts of the world have traditionally treated nomads as "a social problem," while greatly underestimating the nomads' economic contributions.[35] Joseph Ginat, another Israeli anthropologist with extensive fieldwork experience among the Bedouin, offered a broad, but generally true, observation: "The authorities in all Middle Eastern countries are always interested in having more control over the population and in diminishing the extent of nomadism."[36]

As in the Arab Middle East, so in Central Eurasia, nomads have employed various strategies to evade state authority or to rework the intended results of state planning agencies. It is nonetheless undeniable that pasture use has been subject to different regimes of authority throughout Mongolia's history. Within the geographic realm of Central Eurasia, Mongolia's case is not unique. In fact, some of the most interesting literature in recent years has been produced by researchers of Tibetan pastoral herders. Ken Bauer perhaps best introduces the question of land use and state authority in an article on pasture boundaries in central Tibet. He has argued "that common property regimes are contingent on political processes and that state entities have played a central and abiding role in the delineation of pastures' boundaries, access to rangeland resources, and the mediation of conflict."[37] This is a complex area of inquiry that combines common property theory, information from historical documents (where available) and on-site observations (and interviews of herders when possible).

In very recent times, it has become crucial to measure the weight that the voices of the herders themselves carry in the ongoing debates on land privatization and land use. Mongolia is a new democracy, but Mongolia's leaders often betray the deeply ingrained habits of a one-party political system that lasted for almost 70 years until 1990. This is not just the observation of one foreign historian. Many Mongols would agree with the comment by a Mongolian editorialist who chided his countrymen for not being able to "leave our socialist heritage" including the tendency of socialist political leaders (both Soviet and Mongolian) to reward themselves monetarily and also figuratively with honorary titles and medals.[38]

It is very difficult to gauge public opinion in Mongolia, and even more so among the mobile population of herders. Buried within all the fieldwork studies and reports by international aid agencies, there are inklings of what Mongolia's herders may favor along the spectrum from preserving open access to pastureland to introducing private landholding, fencing, and so on. It is revealing that the Open Society Forum's 2004 public perception survey found an overwhelmingly positive response among herders to the idea that pastureland should be used as common land; less than 10 percent of the herders interviewed favored "individual ownership" of pastureland.[39]

In terms of their involvement in democratic processes, Mongolia's herders turn out to vote in impressive numbers. In June 1996, while traveling in Bulgan in South Gobi (Ömnögov') Aimag, I noted that even when transportation to and from a voting station was not convenient, herders would find a way—and if not, the mobile ballot box would find them. A motorcycle conveying a ballot box was still making the rounds to distant locales at dusk on a June evening in the Gobi.

Linda Gardelle estimates an average 80 percent voter participation rate among herders, but she notes that herders' political participation in *bag khurals* (district assemblies)—the most local of councils—is far more common than their participation in, or election to, aimag khurals. Not surprisingly, it is simply not feasible for a herder to leave behind his or her livestock to campaign for a seat in the *Ikh Khural* (parliament) in Ulaanbaatar.[40]

In understanding the high levels of political awareness and participation among Mongolian voters, both rural and urban, it is important to credit the socialist legacy. Because political participation was mandated in socialist-era Mongolia, a culture of political participation preexisted the advent of multiparty democracy in the 1990s. Although casting votes in socialist Mongolia, as in the USSR, was meaningless in one sense, given the absence of competing candidates on any given ballot, going through the motions of voting and serving on local councils ingrained a familiarity with democratic practices—ironically even in predemocratic Mongolia. Since 1990, elections have indeed become meaningful, and herders have participated in droves.

Herders thus exercise their democratic rights in voting for members of parliament and for prime minister; the prime minister appoints the aimag governors, who in turn have been nominated by aimag khurals (assemblies that consist of elected members). But how do herders obtain information about candidates running for these various elected positions? Much information comes from the televisions that have become quite common in herders' *gers* (felt tents). The electricity powering televisions and receiver dishes comes from solar panels or windmills that are installed adjacent to herders' gers. Thus, the rather romantic image of the isolated Mongolian herding household is nowadays belied by the sight of such modern accoutrements that allow information about domestic politics and the wider world to filter in.

Even in a governmental structure based on equal representation, one wonders how closely Mongolian herders' priorities are taken into account and acted upon by the elected officials sitting in Ulaanbaatar. Does the vote of the urban resident outweigh that of the rural herder? Herders' priorities are being challenged by other growing sectors, including mining, tourism, and sedentary agriculture, all of which have strong proponents and well-financed advocates. Chapter 6 will outline the challenges from these other sectors and examine the compatibility of pastoral nomadic herding with other land uses.

Figure 1.1 Herders' ger with satellite dish and solar panel, Zavkhan Aimag, summer 2005.

Figure 1.2 South Gobi gers with satellite dish and windmill, summer 2010.

The role of agriculture among the Mongols is a key issue in the history of land use. The fact that agriculture has sporadically been practiced by the Mongols throughout much of their history may surprise readers. It was certainly the case that many a traveler, like the eighteenth-century Scottish doctor John Bell, could cross the Mongolian steppes without seeing any evidence of agriculture. Bell, traveling with a Russian embassy sent by Peter the Great to the court of the Qing Emperor Kangxi in the years 1719–1722, came away with the impression that "these people [the Mongols] do not trouble themselves with ploughing, or digging the ground in any fashion; but are content with the produce of their flocks."[41] This is a stereotype that has resisted attempts at deconstruction. Foreign travelers generally have reduced the Mongolian diet to one of extreme animal-based simplicity, ignoring the fact that Mongolian nomads historically have traded for grains or cultivated some themselves. Roy Chapman Andrews, the American explorer and naturalist, offered this characterization in 1921: "The Mongol's food consists almost entirely of mutton, cheese, and tea."[42]

It may be challenging to imagine agricultural practices in such a harsh, arid climate. Yet, the Mongols themselves have embraced the idea that agriculture will play an essential role in their future economic well-being. In the spring of 2008, for instance, with food costs rising dramatically in Mongolia as in the rest of the world, Mongolian citizens took to the streets of Ulaanbaatar to protest a 50 percent rise in the cost of a loaf of bread.[43] While wheat exports from Russia helped somewhat to alleviate the situation, the Mongolian Ministry of Food and Agriculture announced a new program "to turn virgin lands into arable land for farming."[44] As we shall see, the idea of opening "virgin lands" is a recycling of a socialist-era emphasis on self-sufficiency in agriculture as a worthy goal. In recent years, Mongolian government spending on agriculture has eclipsed government spending on herding, reflecting this reorientation of policy priorities.[45]

As demands for greater agricultural self-sufficiency grow in Mongolia, pressure on the land comes from yet another source: mineral and energy exploration and exploitation. As of May 2008, 28 percent of all land in Mongolia had been locked up under license to mining companies for exploration and exploitation.[46] Of all the licenses that had been issued for mining, 44 percent were for gold mining, with fluorspar and iron following. The evolving and often ambiguous roles of aimag and sum governors in allocating land and in resolving land use conflicts among the competing interests (herders, mining companies, tourist companies) are crucial to an understanding of land use nowadays.

Of concern also is the fact that China heads the list of foreign countries with the most numerous mining and exploration licenses in Mongolia.[47] Land use issues are not only limited to domestic policies and priorities within Mongolia, but also have clear foreign policy ramifications. With the most populous nation on earth as its neighbor, Mongolia will continue to be coveted for its very desirable natural resources—not only minerals, but also oil and natural gas. Economic exploitation through control of land and mineral rights presents a more genuine threat than military action by China, which would be highly unlikely, but not entirely unimaginable. China's role as overseas resource exploiter has been highlighted by its activities in Africa, where it is pursuing both underground resources (minerals, oil) and also aboveground wealth—farmland. China has devoted five billion dollars to agricultural projects in Africa, with an emphasis on rice production. Because of climate, desertification, and other variables, Mongolia has escaped the threat of agricultural development by its populous neighbor, yet the Chinese are heavily invested in exploiting Mongolia's underground mineral wealth.[48]

As foreign investors navigate the ever-changing currents of Mongolian regulations and laws pertaining to mining, they face a very basic question: must the labor force involved in mining and other resource extraction be Mongolian in composition? At the enormous Oyu Tolgoi copper- and gold-mining project in the South Gobi, operated by Canadian-owned Ivanhoe Mines, at least 60 percent of the project's employees, or roughly 3,900 workers, are supposed to be Mongols as set by the Investment Agreement.[49] A May 2010 report by the minister of finance identified only one thousand or so Mongols as qualified to pave roads and build utility lines in support of the mining project. The minister suggested importing 2,600 Chinese workers to fill the shortfall of qualified workers. This proposed solution was reportedly not met with enthusiasm by members of the government's Standing Committee on Foreign Policy and Security. A counterproposal suggested hiring less-qualified Mongols to receive on-the-job training, rather than open the floodgates to Chinese laborers. By mid-July 2010, a reported 2,250 Mongolian employees were working at Oyu Tolgoi, a number representing 70 percent of the workforce; there was no assessment of the skill level of these employees.[50] The Mongolian Ministry of Education signed a memorandum of mutual understanding with Oyu Tolgoi in July 2010 that would enable 12 colleges and vocational centers in Mongolia to train future workers headed to Oyu Tolgoi.[51] Emotions flared up again when it was announced in May 2011 that 6,949 Chinese workers would be hired to work at Oyu Tolgoi. With high unemployment rates in Mongolia, the decision to import even more Chinese workers was extremely unpopular.[52]

The contentious proposals to import Chinese workers and miners into Mongolia had historical precedent. In 1896, a Russian baron, Victor von Grotte (also spelled von Grot), began operating seven gold mines in northern Mongolia; he obtained the mining concessions from Mongolian princes, and he employed Chinese labor. The British authors of the 1914 work, *With the Russians in Mongolia*, observed "very large numbers of Chinese . . . employed in the works supervised by Russians."[53] It was thought that mining was distasteful to the Mongols: "The only gold-mining concern, the Mongolore Mining Company, started by Mr. Victor von Grotte, to develop the enormous mineral wealth in North Mongolia, was obliged to import Chinese labour from 1000 miles away, simply because the Mongols refused at any price to work the mines."[54] The British observers, writing in 1914, ascribed what they perceived as the laziness of the Mongols to "nomad habits."[55]

In addition to exhibiting a profound misreading of pastoral nomadic work habits, the authors of *With the Russians in Mongolia* also overlooked a traditional aversion among the Mongols to disturbing the earth (*Etügen Eke* or "Mother Earth"). In the mid-twentieth century, as Tsogt-Ochiriin Lookhuuz, the former director of state farms in the socialist era recalled, he had to overcome a deep-seated resistance among herders to disturbing the soil by plowing in order to plant crops.[56] In more recent times, however, dire economic circumstances have driven tens of thousands of Mongols into small-scale artisanal mining for gold; these so-called ninja miners labor in unsafe conditions to dredge up tiny amounts of gold to supplement meager incomes.

In the socialist era, before the Sino-Soviet split resulted in the expulsion of most of the Chinese labor force from Mongolia in 1983, Chinese workers had, to their credit, constructed several of Ulaanbaatar's apartment buildings, bridges, and other infrastructure. Mongols in the 1970s and 1980s would say privately that Chinese-constructed apartment buildings were far superior to their Soviet-constructed counterparts; this assessment, may, however, have reflected the hostility that many Mongols felt toward their Soviet "elder brother." Nowadays Mongols tend to denigrate Chinese construction as inferior, again reflecting current attitudes toward the many Chinese workers and businessmen who have flooded into Mongolia since 1990.

Unfortunately, the attitude of cultural superiority often displayed by skilled Chinese workers has only worsened a simmering anti-Chinese sentiment that has long been evident among the Mongols. In early July 2005, during a brief stop at the hydroelectric dam under construction on the Chono Kharaikh River in Khovd Aimag in western Mongolia, I inquired of the Chinese boss whether he had any Mongols on his crew. His comments were made with undisguised condescension: of the 60 workers employed constructing the dam, all were Chinese; none were Mongols. When I asked why he had not hired any Mongolian workers, he replied that there was a problem with Mongolian labor—in his words, they couldn't even follow simple instructions.

Thus, the issue is not only at the level of foreign investment and ownership, but also at the ground level of an actual workforce. What ownership can the Mongolian people feel if the workers extracting ores from underground or building infrastructure projects above ground are Chinese or other foreign nationals?

As Chinese and other foreign investment increases in Mongolia, it is legitimate to ask whether Mongolia's elected officials will be able to avoid the temptations of amassing great personal wealth through ties with foreign-controlled mineral companies. Or, will Mongolia's politicians find the Russian model of a corruption-laced, autocratic oligarchy the easiest model to emulate?

There are genuine causes for concern. In a 2009 survey on corruption, compiled by the Asia Foundation and the Sant Maral Foundation, the overall trend pointed to Mongolian households paying significantly increasing amounts in the form of bribes from 2007 to 2009.[57]

The Land Utilization Office earned top ranking as the most corrupt government agency in all eight surveys conducted from March 2006 through September 2009. This does not bode well for the adjudication of land issues in the future. The categories of customs, mining, and judges claimed most of the second and third place ranks under "areas considered as most corrupt" in the same three and a half year period.

In October 2010, it was reported that Mongolia's Anti-Corruption Authority (ACA) conducted a survey among private enterprises. The survey results were not encouraging: among private enterprises, 81.2 percent of respondents believed that Mongolia was rife with corruption; 27 percent had experienced corruption at the hands of federal government organizations; 32.4 percent had faced corruption in capital and provincial government organs; and 40.7 percent had endured corruption in the private sector.[58] Among newly emerging forms of corruption are the use of paid middlemen by applicants seeking jobs at mining companies and questionnaires by at least one mining company requiring vocational training program applicants to state political party affiliation and information on any relatives who work for the government or for international aid organizations.[59] This survey suggests that a culture of corruption has taken root in Mongolia, and that mining companies, foreign economic interests, and wealthy Mongols will exert increasing control over the substantial (but finite) resources of the country.

How Mongolia's land is managed in the years to come will be a key factor in the country's economic and social development. Livestock herding will continue to play a significant role across the Mongolian countryside. According to a 2009 report, about 80 percent of Mongolia's land is devoted to livestock herding by pastoral nomads.[60] The category of "agriculture" provides roughly a quarter of Mongolia's Gross Domestic Product (GDP), and livestock herding accounts for 80 percent of so-called agriculture. Among Mongolia's 2.8 million citizens, roughly 180,000 households or one-third of the population are engaged in some aspect of livestock herding. Mongolia's pastoral nomadic herders should be included in the national debates over land use; their opinions should be given careful consideration. Crop agriculture, mining, and tourism all have strong proponents, varying degrees of governmental support, and, again to differing degrees, foreign and domestic investors. By comparison, livestock herding, while trumpeted as integral to Mongolia's pastoral nomadic history and culture, is often not at the forefront of politicians' concerns or agendas.

If livestock herders, currently held responsible for pasture deterioration through overgrazing, embrace sustainable approaches to land use, they stand a good chance of preserving their way of life and thriving economically. The task of finding an environmental and economic balance among the competing interests in land use ideally will be taken up by the government of Mongolia. In a more activist role, the government will need to aspire to fairness, transparency, and the will to shed its own corrupt habits. As in earlier periods of Mongolian history, but even more so now, herders will ultimately have the choice of either actively working with external sources of authority (in this case, their elected officials) in determining their fate or working around (ignoring) unpopular policy initiatives if their voices are not heard.

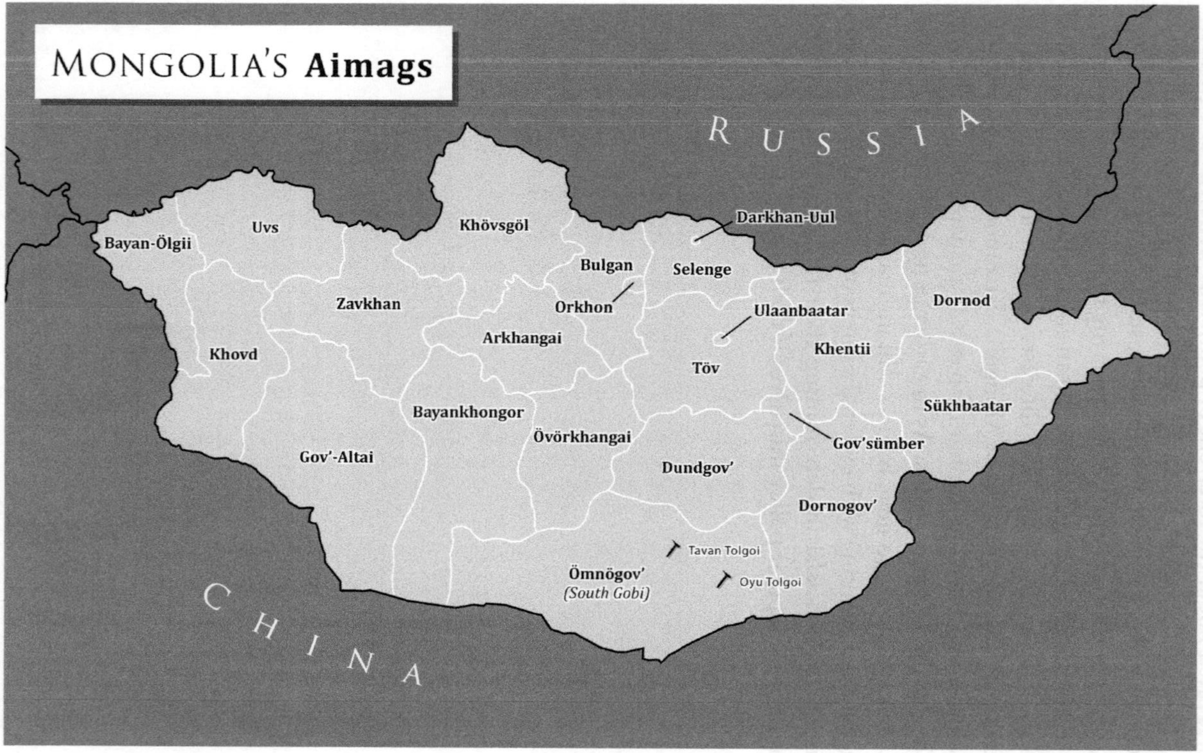

Map 1 Mongolia's Aimags

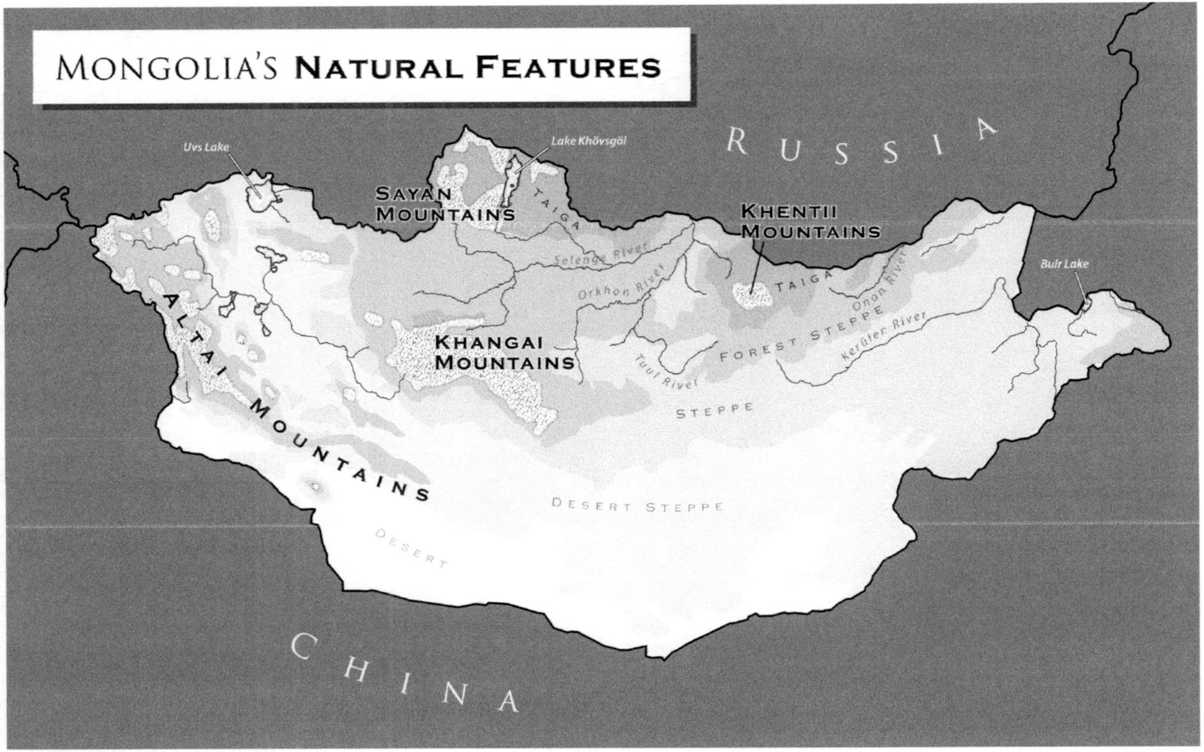

Map 2 Mongolia's Natural Features

CHAPTER 2

MONGOLIA'S REGIONS: LESSONS IN VARIATION

MONGOLIA'S ECOLOGY AND ENVIRONMENT LEND THEMSELVES to different, yet overlapping, categories. For instance, one may define Mongolia's regions by mountain chains: the Khangai region (central-western Mongolia); the Khentii region (central-northeastern Mongolia); the Altai region (south-southwest); and the Sayan region (along the northern border with Russia). Or, one may instead use the convenient distinctions of vegetation zones: taiga in the north; mountain-forest-steppe (north-central); steppe (central); and desert-steppe and desert (southern tier). In terms of livestock herding, Mongolia's pasturelands may be divided into five zones: the Khangai-Khövsgöl forested mountain region of the northwest; the Selenge-Onon Rivers region of north-central Mongolia; the high Altai Mountain zone; the central-eastern steppe lands; and the desert and steppe region of the Gobi. Each of these five zones supports different mixtures of herd animals.[1] Whichever geographic categorization one uses, all these ecosystems overlap, but generally run as lateral bands across a map of Mongolia.

Distinctions in geography, vegetation, access to water, and climate all affect how nomads use the land. These distinctions also make generalizations a bit difficult when we look more closely at widely divergent patterns of nomadic migration, the varied composition of herds, and the degree to which herders may or may not be connected to larger economic and marketing networks. Rather than homogenize significant variations, this chapter will explore the diversity of Mongolia's natural world and the accommodations that humans have had to make to exist in oftentimes difficult circumstances.

THE GOBI: EXCEPTIONALISM AND EXTREMES

For many people, the Gobi is so tightly interwoven with the image of the "real Mongolia" that it may seem odd to refer to Gobi exceptionalism. Yet, everything about life in the Gobi—for humans as well as for herd animals—is defined by extremes and exceptions. While the pastoral nomadic way of life encounters enormous challenges throughout Mongolia, nowhere is it so demanding as in the cold desert environment of the Gobi.

The Gobi desert in its entirety encompasses some 390,000 square miles, constituting about one-third of Mongolia's southern regions as well as extended regions of Inner Mongolia in China's north. It is a pebbly, rock-strewn desert with sparse but nutritious vegetation for herd animals. Only about 3 percent of the Gobi in Mongolia consists of sand, yet its tall, windswept sand dunes covering an area of over 900 kilometers at Khongoryn Els in Gobi Gurvansaikhan National Park invariably leave stunning impressions upon visitors.[2]

As a cold northerly desert, the Gobi's harshness is felt most in winter when temperatures may plunge to –40 degrees Fahrenheit and are often accompanied by ferocious northerly winds. Fall and spring in the Gobi also witness great winds with dust storms common in the spring months. Sands from the Gobi end up clouding the skies over Beijing in March and April, and contribute to respiratory problems.

In summertime, the challenge is finding water, with annual precipitation averaging less than four inches. Shallow wells at only eight- to ten-feet deep historically have allowed for extensive grazing of herd animals in Gobi regions, although migrations from pasture to pasture were and continue to be of necessity more numerous in the Gobi than elsewhere in Mongolia. In postsocialist Mongolia, however, water availability and access have become critical issues. After the collapse of the socialist-era *negdel* ("socialist collective") system, most wells constructed in that period fell into disrepair through neglect and vandalism. Many of these wells were drilled to depths ranging from 30 to over 200 feet, and they operated with mechanical pumps. With the privatization of negdel assets in the early 1990s, ownership rights over wells were generally undefined; lack of clear possession led to broken well pumps being left unrepaired or sold for scrap metal to Chinese buyers. With spare parts for well pumps no longer being supplied by the Soviet Union, only one in five wells in Mongolia functioned in 2004 out of some 24,600 that had been built nationwide in the previous four decades. While the socialist era saw the expansion of herding in the Gobi owing to the drilling of more wells, since the early 1990s herders in the Gobi regions have withdrawn from rangelands that had once been used.[3]

Gobi herders not only face ongoing problems with access to water, but also the potability of available water is often suspect. Gold mining in many regions of Mongolia has in recent years rendered the water of countless streams, rivers, lakes, and underground sources unsafe owing to the use of mercury that, despite being illegal, continues to be used in artisanal gold extraction. With an estimated 31 percent of Mongolia's total territory as of 2009 under license for mining exploration and exploitation, the outlook for herders in terms of both pasture and water access remains troubling.[4] The October 2009 signing by the government of Mongolia of a long-term arrangement with Ivanhoe Mines to construct and operate an enormous copper and gold mining complex in the South Gobi at Oyu Tolgoi includes much language assuring compensation to herders whose resources are impacted adversely as well as guarantees of access to sufficient water for livestock.[5] There is considerable doubt concerning the government of Mongolia's willingness to conduct the necessary environmental oversight at this project; at least some elected officials may prove unable to resist the lure of amassing private fortunes in their own lifetimes through the potentially lucrative mining sector.

Plans are being developed to ensure that Mongolia's citizens will share in the country's new mineral-derived wealth. In 2010, the government of Mongolia established a 100 percent state-owned company under the name "Erdenes MGL" for the South Gobi Tavan Tolgoi coal deposit that has an estimated reserve of 6.4 billion tons of coal.[6] According to a draft resolution submitted to the parliament, shares of Tavan Tolgoi will be distributed to each and every Mongolian citizen, a plan that is reminiscent of the State of Alaska's Permanent Fund, which is derived mainly from oil revenues that pays out dividends annually to all residents of the state. For the foreseeable future, mining and herding will continue to overlap geographically across the Mongolian landscape. Difficult choices loom in the environmental, economic, and cultural spheres.

The nutritious natural forage of the Gobi traditionally has supported all of Mongolia's so-called five animals (*tabun khoshighun mal*)—sheep, goats, camel, horses, and cows (a category that includes yaks and *khainag* or yak-cow hybrids). Cows, however, have been and continue to be the least numerous of the five animals on account of the difficulty they encounter grazing in the short and sparse Gobi grasses. The number of goats, on the other extreme, has grown to unprecedented levels since the introduction of free market reforms in the early 1990s in response to the global market for cashmere, the fibers from a goat's undercoat. Mongolia's goats provide 30 percent of the world's cashmere. The staggering growth in numbers of goats—up to 18 million according to the 2007 livestock census[7] with the majority in the Gobi aimags—continues to strain the Gobi's pasture lands and to contribute to overgrazing and desertification in this delicate ecosystem. In addition to cashmere, other animal products such as meat, dairy, and wool (from camel and from sheep) all provide income and sustenance for Gobi herders.

Because of the harsh climate, limited access to water, and sparse vegetation, herding patterns in the Gobi have traditionally differed in significant ways from patterns observed elsewhere in Mongolia, that is, in the steppe and northern forested regions. To understand the peculiar characteristics of nomadic herding and land use in the Gobi, one cannot find a better guide than the writings of Andrei Dmitrievich Simukov (1902–1942?), a renowned scholar of virtually all aspects of Mongolian life. Simukov, whose studies of the Mongolian countryside established him as the preeminent specialist on Mongolian geography and ethnography, was the leader of 15 large-scale expeditions and numerous lesser excursions in Mongolia from 1927 to 1939.[8] His daughter, Nataliia Simukova, has estimated that Simukov's travels extended to practically all areas of Mongolia, and that his itineraries traversed some 70,000 kilometers.

Simukov set out on his first Gobi expedition in July 1927, and the expedition was in the field for five months. Within the next ten years

Simukov led four more Gobi expeditions (1930, 1931, 1932, and a winter expedition in 1937). In addition to making invaluable contributions to the fields of natural science, geography, and cartography, Simukov increasingly directed his research toward economic geography starting in the early 1930s. Among his voluminous and highly detailed works on Mongolia's pasturelands is a 200-plus-page work entitled "The Gobi Pastures of the Mongolian People's Republic," (in Russian) published in 1935–1936.[9]

Simukov's many observations into the human landscape of the Gobi and other regions of Mongolia present us with unique benchmarks for understanding the nomadic way of life in the Mongolian countryside of the 1930s and the Mongolia of our own times. It is also important to underscore the fact that Simukov was able to travel widely and make field notes on the pastoral nomadic way of life before Mongolia's herders were forced into socialist collectives (negdels). With the demise of the negdels in the early 1990s, the lives of Mongolian herders today may bear an eerie resemblance to the lives of their grandparents and great grandparents. The following pages will present Simukov's findings and will suggest that many of his ideas and conclusions are still pertinent to Mongolia today.

Some of Simukov's observations on the pastoral nomadic way of life centered on the similarities between the Gobi and other regions of Mongolia; others underscored the extremes and particularities of the Gobi. Many of Simukov's interregional comparisons serve to contrast the Gobi with the Khangai region. Of course, there are also some very significant variations among the other non-Gobi regions of Mongolia that merit attention.

The Gobi Contrasted with the Khangai and Other Regions of Mongolia

Among the primary contrasts between the Khangai region and the Gobi region are the number of nomadic migrations and the distances covered in each migration. A recent study describes the Gobi as the region where nomadic herders move the longest distance as well as the most frequently in all of Mongolia, from 20 to 30 kilometers every 15–25 days.[10] Another researcher who has spent considerable time among Khangai Mountain herders describes four migrations a year, corresponding to the four seasons.[11] Spring and autumn encampments are typically used for a month or so, while winter and summer sites are longer term. Because the Khangai Mountain region of Mongolia has reliable pastures and water, four migrations suffice, but in drier areas of the country, like the Gobi, more numerous migrations are common. Interestingly, Simukov in his time observed Gobi herders moving camp less often, but covering more distance than did the Khangai herders when they moved camp.[12]

Simukov's observations are highly suggestive of the ways in which the natural environment may shape the structure of family and community life in Mongolian society. The scarcity of feed for herd animals in the Gobi, noted Simukov, forces the population to spread out in as scattered a fashion as possible. Herders' campsites in the Gobi, in Simukov's time as today, do not feature several households and *gers* in one place. Most Gobi families nomadize alone, not with other herder families. The Gobi thus differs from many areas of central-northern Mongolia where campsites are well populated by several gers. Simukov encountered one or two gers, or on occasion three, camped near a well or a spring in the Gobi; only in atypical spots in the Gobi where water and pasture were more plentiful did he find a larger gathering of gers, but even in these cases, the gers were spread out individually or in pairs throughout the area. In areas lacking good access to water, *ayils* (encampments of gers) would be quite remote from one another. Simukov noted that the spacing of ayils at great distance from one another led to an insular and monotonous way of life, punctuated only by occasional trips to a nearby monastery or trips for trade purposes.

Interestingly, the Gobi pattern of pastureland usage was not as predictably regular as in the mountainous regions of north-central Mongolia, including the Khangai. Winter and summer pastures were intermingled in the Gobi, though in general Gobi herders preferred open spaces in summer and mountainous or hilly terrain in winter. Simukov concisely described the simplicity of wind-protected winter livestock encampments in the Gobi in his era: "The sole structure present at a winter encampment is a low semi-circular enclosure, open from one side, and made of rocks, intended for sheep and goats."[13] Today, however, there are many more structures on the landscape in the Gobi *aimags* as well as in other aimags throughout Mongolia. It is far more common nowadays to spot permanent storage sheds and outbuildings standing adjacent to winter livestock enclosures in the Gobi.

In the voluminous, data-laden study of Gobi pastures that he wrote in 1935–1936 (alluded to earlier in this chapter), Simukov estimated that in the Gobi zones that he had studied, almost 90 percent of herder households nomadized along a 25-kilometer diameter.[14] The choices of vegetation, from a herder's vantage point, would have been far more varied than one might suspect, given the apparent monotony of the landscape. Simukov offered a detailed typology of Gobi pastures, classifying seven distinct categories based on vegetation: grassy desert-steppe; small shrubbery (saltwort) desert-steppe; *toirom* (Mongolian salt marsh without vegetation) that is hilly and sandy; large-shrub desert; sand dunes stabilized by vegetation; a mountainous variant of grassy desert-steppe; and *deresü* (salt-marsh meadows).[15]

According to Simukov, Gobi pastures are generally characterized by the high quality of grasses and shrubs and by the sharp response to precipitation. Given the extreme unevenness of precipitation, the Gobi fodder base tends to fluctuate dramatically. In drought conditions, nomadic migrations may become disorderly and even chaotic. In 1926, Simukov witnessed drought-stricken Gobi pastures and recorded his anguished reaction:

> I witnessed (Khan-kökchün-ula, 1926) abandoned campsites, surrounded by heaps of dead sheep; dogs, wolves, vultures, and eagles could not even keep up with these. Livestock that remained alive wandered dejectedly, barely moving their legs along the exposed rocky expanse, searching for withered patches of last year's grasses. The camels literally swayed in the wind. Yesterday's man of wealth becomes a pauper.[16]

While drought in the Gobi regions is frequent, in Simukov's observations, he found that it rarely encompassed the entire Gobi region all at once. In most cases, the worst affected and the most barren pastures would be interspersed with relatively normal ones. Based on his interviews with local inhabitants in the South Gobi, Simukov estimated that a large-scale drought generally occurred every 25–30 years, while smaller, more localized droughts occurred once every 4–5 years.

In spite of the ever-looming threat of drought, the Gobi landscape in Simukov's time as well as nowadays supports a variety of herd animals. Simukov described the camel as the chief herd animal in the Gobi, and he estimated that 64 percent of all camels in the Mongolian People's Republic (MPR) were pastured on Gobi lands. In Simukov's estimation, goats constituted only 11 percent of Gobi herd animals, and only 7 percent of nationwide herd animals.[17] The camel no longer occupies the top of the Gobi herd hierarchy in terms of numbers, as today the goat population far exceeds the camel population. As noted earlier, the international market for cashmere has led to an enormous expansion nationwide of the goat herd.

The drop-off in camel population between Simukov's era (1920s–1930s) and the present began in the negdel era, when Mongolia's socialist collectives, heavily subsidized by the Soviet Union, were able to supply trucks to herders for their periodic nomadic migrations. While trucks thus replaced camels for transport needs in the socialist era, in the early years of the postsocialist-era herders faced a lack of subsidized mechanized transport as well as high fuel costs. As a result, a renewed reliance upon the now less plentiful camels occurred. In 2010, the continuing drop in camel numbers provoked news headlines in the Mongolian press like, "In 20 years, there may be no camels."[18] Certainly, the steep decline in Bactrian (two-hump) camel numbers is alarming: in 1954, camel numbers peaked at 896,500, but by 2010, only 270,000 remained in Mongolia. The *zud* of winter and spring 2009–2010 led to the death of more than 10,000 camels. With camels giving birth to a single offspring only once every two years, the current rate of reduction has understandably led to greater efforts in Mongolia to conserve the remaining population. Camels are the only category of Mongolia's five domestic herd animals to have experienced a long-term, steady decline in numbers, though over the past decade this decline may have at least slowed.[19]

As for the bovine population, yak are herded in the higher elevations of both the Altai Mountains, which span the central and western Gobi aimags, and the Khangai Mountains. Simukov noted that mainly khainag were raised in the Gobi. Cattle simply cannot graze easily among the short, sparse Gobi grasses, and they are generally found more in the peri-Gobi zone, not in the heart of the Gobi.

The mountainous type of Gobi pasture is well suited for sheep and goats since they have little difficulty with steep terrain. Horses, sheep, and goats generally thrive upon the plant composition of the mountainous pastures. Camels, however, prefer lower zones where the valley vegetation is more to their liking. Cattle are ill suited to mountainous Gobi pasture, though yak do well. Because camels prefer grazing on the flats where salt marshes are located, wealthier Gobi herding households would, according to Simukov, separate into two parts: one ger would nomadize with the camels on the plain, while the other(s) would nomadize in the mountains with the sheep, goats, and horses.[20]

Overall, Simukov in several of his field reports extolled the high quality of Gobi grasses as key to the abundance of both herd animals and wild animals in the Gobi. The meat of sheep from Gobi pastures, according to Simukov, was highly aromatic, and he noted "the great paradox in Mongolia that among the rich, lush, varied grasses of the Khangai, livestock prefer formations that are closest to Gobi [vegetation]."[21] There seems to be a correlation between sandy soil and high-quality sheep meat (mutton). While traveling in Zavkhan Aimag in the summer of 2005, I spoke with the then governor of Telmen Sum, Dolgorzhav, who claimed that the sandy soil in his *sum* produced some of the finest-tasting mutton in all of Mongolia.

Perhaps one of the greatest regional differences between the Gobi zones and the other regions of Mongolia pertains to haymaking. In the 1920s—1930s, Simukov observed that haymaking was feasible on a small scale in the Central Khangai region. Yet, even in the Khangai, haying might not be worth the effort in some cases because alpine fields could be hayed only when in bloom, but in summer, abundant rains tend to fall on the higher areas, making it very difficult for the hay to dry. Nonetheless, Simukov urged local populations to learn how to produce hay as an essential factor for the survival of the pastoral nomadic way of life.[22] Echoing Simukov's observations, J. M Suttie, in his 2005 work on "Grazing Management in Mongolia," noted that in most of Mongolia natural herbage is at its best in the latter half of August, a time of potentially heavy rains.

Nowadays efficient haymaking in the short time span that is feasible continues to be hindered by insufficient mechanization and the need for extra labor.[23] Herders in the Khangai Mountains use scythes to cut the long grasses at their reserve pastures, and they store the hay on the roofs of their winter/spring livestock shelters.[24] Nonetheless, during the summer of 2010, I noted far more evidence of small tractor use in both South Gobi Aimag and, in particular, in Selenge Aimag where Hyundai tractors were quite numerous.

Haymaking demands a coordinating of labor and the assignment of specific pastures to be devoted to hay production. In Khan-Undur Sum, which is not far from Tsetserlig, the capital of Arkhangai Aimag, Simukov noted in 1929 that haymaking was supported by the Tsetserlig administration. The dense, high grasses in this sum made excellent hay, and Simukov witnessed disputes over the better hayfields. Predicting that intervention by the authorities would become necessary in the distribution of hayfields, Simukov foresaw this as "the first step towards a land management system and a transition to a semi-settled way of life."[25]

In the socialist era, the negdel leadership regulated hayfields and hay cutting, and the lines of authority were drawn with clarity. Nowadays, however, the absence of structured authority vis-à-vis hayfields and other pastureland presents problems that are not easily resolved. As Ole Bruun in his study of Khotont in southeastern Arkhangai in the 1990s discovered, the collapse of socialism made hay cutting and transport very difficult. Bruun reported that most herders let their animals graze throughout the winter with only enough hay in storage to see the animals through short emergencies, that is, heavy snow cover on pastures.[26] Another source reports that in the past-1990 years, hay from natural pasture produced the only available fodder in Arkhangai Aimag. Hay pastures were unregulated, the yields were very low, and the cutting of hay was described as "competitive."[27]

Over the past decade, international nongovernmental organizations (NGOs) have become involved in facilitating hay production in small-scale, pilot programs. A prime example is the "Green Gold" Pasture Ecosystem Management Program, under the auspices of the Swiss Agency for Development and Cooperation. In the quest to make the Mongolian livestock industry more productive and to make pastureland sustainable in the long run, the "Green Gold" project has sponsored fencing and irrigating of hayfields as well as the digging and reclaiming of wells, and the improvement of degraded pastures in 27 sums across Mongolia from 2004 to 2009.[28]

In addition to natural pastureland haying, there is a history of sown fodder in Mongolia. In the socialist era, the negdels and state farms planted and cultivated hay crops. Conditions in the western aimags particularly favor the cultivation and irrigation of hay crops.[29]

Yet, even with the hurdles presented by climate, economic resources (or the lack thereof), and the uncertainties of pooling labor, haymaking is clearly viable in many parts of Mongolia. In his travels through Mongolia, Simukov noted other areas where haymaking was in progress—along the Selenge River and in the Ider River valley in the north, for instance.[30] But what about the Gobi? Simukov does not mention hay fodder in the Gobi in his era. In the socialist era, the Mongolian government subsidized the annual transport of hay and fodder to South Gobi Aimag in amounts ranging from 30,000 to 40,000 tons a year. Most of this hay and fodder came from a distance of 1,000 kilometers to reach needy Gobi collectives.[31] Attempts at growing yellow-flowered alfalfa in smaller irrigated areas of the Gobi met with some success in the socialist era, but those areas are now devoted to growing melons and vegetables that earn greater income.[32] The Gobi simply does not have natural hay fields, given its sparse grasses. Recent foreign aid has been devoted more to "risk management technologies" in areas of the Gobi that have suffered from zud and drought in the past. The Gobi Forage Project under United States Agency for International Development (USAID), for instance, operating from 2004 to 2008, aimed to monitor forage availability and to provide early warning in dire winter conditions in six Gobi aimags.[33]

It is worth emphasizing the hardiness of Mongolia's indigenous livestock breeds. With limited hay production and storage, Mongolia's livestock relies on dried grasses on winter and spring pastures for some 180–200 days of the year. All Mongolian breeds of livestock to varying degrees are capable of pawing through thin layers of snow to graze on dried vegetation beneath. As one Mongolian scholar writes, "this is what makes Mongolian livestock so biologically remarkable and valuable."[34] Mongolian livestock has adapted to the extremes of heat and cold that regularly occur by season. Indeed, the adaptability of the livestock to varying conditions and their careful tending by herders who pass their specialized knowledge of pasturage, breeds, and so on from generation to generation combine to promise the long-term survivability of this way of life.

FUEL AND GARBAGE

In most of Mongolia, herders rely to varying degrees on *argal* (dried dung from livestock, especially from bovines and camels), as fuel for woodstoves in gers. Argal burns efficiently and without odor; it serves

Figure 2.1 Cooking on the ger stove in the South Gobi, 1996.

to heat the ger and to provide fuel for cooking. The use of dried animal dung is an environmentally sound methodology practiced by pastoral nomads past and present throughout Central Eurasia. Before the advent of woodstoves, introduced by the Russians into Mongolia in the socialist era, an open fire pit in the center of the ger was also typically fueled by argal (see figure 2.1).

The collection of argal is generally women's and children's work. A special basket and pitchfork enable the collector to scoop the dried animal dung up and fling it in the basket, worn with straps on the collector's back, eliminating the need to stoop down (see figure 2.2).

The availability of firewood varies by region in Mongolia, and thus the ratio of argal to wood as fuel to warm the ger varies as well. The Gobi, for instance, differs from other regions in herders' dependence upon shrubs as fuel. In the 1930s, Simukov noted that the best source of fuel in the Gobi region was shrubbery. He described Gobi shrubbery as burning powerfully, throwing heat effectively with little smoke, and having a pleasant aroma. The drawback, of course, was that such shrubs burned quickly—with the exception of saksaul.[35] Saksaul or *Haloxylon ammodendron* (Mongolian *zag*), a woody shrub that can grow up to 12 feet in height and live for several hundred years, is a hardwood that burns for a longer period. Because nowadays the

Figure 2.2 Woman with argal collection basket, Arkhangai Aimag, 2002.

Gobi saksaul forests have become depleted by overharvesting, most saksaul are younger than 25 years; only dead saksaul is permitted to be collected for fuel.[36] Older saksaul help prevent erosion because of their very deep root systems that stabilize sandy soil; mature saksaul also combat wind erosion and thus contribute to the ongoing battle against desertification. In addition to harvesting saksaul for their ger stoves, herders in the Gobi use the foliage of the saksaul as fodder for their livestock. Camel continue to graze in saksaul forests in spite of environmental regulations. A 2004 report found an unsustainable rate of saksaul harvesting in the Gobi, though actual statistics are not available (see figure 2.3).[37]

Figure 2.3 Gobi saksaul, 2006.

In contrast to the Gobi, in northern aimags such as Khövsgöl and Selenge, herding households rely on firewood more than argal for heating and cooking in gers. A World Bank estimate of the regional distribution in Mongolia of wood use for heating purposes suggests that twice as many households in Mongolia's north rely on firewood than in the south.[38] This should not be a surprise, since much of the forestland that comprises 8.1 percent of Mongolia's total landmass is situated in northerly aimags. Primary among the causes of deforestation in Mongolia is the insufficient management of woodlands—including insufficient enforcement of regulations prohibiting illegal logging for firewood and other products. Forest fires and the impingement of livestock upon reforestation areas have also accelerated the deforestation of the country.[39]

In the absence of any reliable statistics on fuel sources in the Mongolian countryside, we may simply conclude that the ratio of argal to firewood largely correlates to the natural environment, which varies considerably from one area of Mongolia to another.

While outsiders may attribute to pastoral nomads in Central Eurasia an inherent environmental consciousness as part of a larger willingness to romanticize the pastoral nomadic way of life, it is clear that in modern Mongolia, garbage disposal has become a very large problem. The issues of insufficient garbage disposal resources in Ulaanbaatar, the almost total absence of recycling, and the overuse of plastic bags are not limited to Mongolia's urban sector. The Mongolian countryside is littered with discarded automobile parts, tires, trash (especially near tourist ger camps), and windblown plastic bags that one encounters in even the most remote areas. In 2011, Ulaanbaatar and other urban areas in Mongolia inaugurated a ban on the sale of thin plastic bags that are commonly sold at supermarkets and other shops, following the precedent of China, which in 2008, banned stores from giving out plastic bags for free.

The problem of garbage disposal is not necessarily just a modern-day problem. Simukov, writing about the Central Khangai in 1929, observed that among the reasons for removal from one pasture to another in the summer months was garbage accumulation. As garbage piled up, flies were attracted in ever-greater numbers, and the herding household would be very motivated to move to a different locale.[40] Simukov was neither the first nor the last observer to find biting insects a problem both for the livestock and for the herders themselves. Herders in summertime will typically look for pasture that is near water—rivers and lakes—but, of course, a variety of insects such as mosquitoes also camp out near water, making life virtually unbearable for animal and human alike. In many areas of Mongolia, a typical strategy is to move the herds higher up in the mountains where stronger winds may deflect some biting insects; in late summer or early autumn when the insect population has been reduced by the first frosts, the herders will then relocate at water sources.

VARIABLES IN NOMADIC MIGRATIONS

Insects are only one contributing factor in the patterns of nomadic migrations in Mongolia's regions. There are a great variety of factors beyond just the natural conditions of a locale that may determine nomads' selection of seasonal campsites. For instance, the specific size and composition of the herd itself, the relative economic well-being of the herding household, and the larger social context of cooperation or competition over resources all may exert influence on the choice of pasture and campsite. It is even difficult to generalize about winter and summer campsites in terms of elevation.

Simukov himself offered a clear example of regional variation: in Arkhangai Aimag, herders summered in lower elevations and in winter moved higher, but in Khovd Aimag in Mongolia's west, the opposite cycle occurred as summer camps were located at higher elevations than winter ones.[41] A more recent report also notes the fact that in the Khangai and Khentii Mountains in northern and central Mongolia, herders in summertime opt for deep valleys and foothills.[42]

Simukov developed a typology of nomadic migrations that explores such contrasts. The Khangai type of nomadism is characterized by summer camps sited at riverside meadows whose pasturage is replenished by the localized moisture. Khangai winter campsites, located at higher elevations (as noted earlier), are placed in high valleys that are shielded from cold winds but well exposed to sunshine. Simukov characterized the "Western type" of nomadic migration (i.e., Khovd and adjacent aimags) as the opposite of the Khangai type: winter camps were higher than summer ones. The steppe type of migration was characterized by the lack of differentiation between winter and summer pastures in location and in elevation.

What Simukov termed the Övörkhangai type of nomadic migration was applicable to a relatively smaller area along the southern outskirts of the Khangai Mountains—present-day Övörkhangai Aimag and the southwestern part of Arkhangai Aimag. As a variant of the western type, the Övörkhangai pattern saw summer encampments in the higher zones of the Khangai Mountains where fresh, running water was plentiful, temperatures were cooler, and pastures were quickly replenished by moisture. In winter in this type of migration pattern, the nomads and their livestock moved to the relatively snow-free Gobi lowlands. Fall and spring saw numerous, shorter-term, intermediate encampments. Not surprisingly, the distance of annual migrations of households in the Övörkhangai category could reach a total of 300–400 kilometers.

Typical of the eastern plains of Mongolia, the so-called Eastern type of nomadic migration was similar to the Övörkhangai type insofar as in winter a portion of the population nomadizes southward to lower depressions with desert-steppe vegetation. In summer, they go north to the Kerülen River and adjacent steppe lands.

The final category in Simukov's typology was the Gobi nomadic migration. In a "normal" or drought-free year, Gobi nomads do not require a very large area for pasturing livestock. This is because of the relative diversity of vegetation in the Gobi. The various types of pasture necessary for herd animals recur with regularity in each valley or depression and in the surrounding heights. Thus, there is no great advantage to undertaking long nomadic migrations. In this feature, Simukov points out, the Gobi type of nomadic migration somewhat resembles the Khangai type. Of course, in case of drought, very long-distance and even chaotic migrations may occur.

In summer, Gobi nomadic encampments are usually in open, flat areas. By contrast, winter encampments are situated in hilly or mountainous terrain. The camel-herders among Gobi nomadic households remain in the open, flat areas for winter, often in thickets of saksaul. In the Gobi Altai, there are exclusively mountain-based households that never nomadize out to the flatlands. Some wealthier Gobi households divide up, with a few herders pasturing the camels on the flats, while the others pasture the sheep, goats, and horses in the mountains.

As in all such attempts at categorization, Simukov reminds us that there are factors not under human control that may alter such nomadic migrations, summertime drought and winter or spring zud primary among them. Certainly, one of the more fascinating variations in herding patterns is the *otor*, a term that refers to long-distance migration of herders with livestock to seek out better pasture on a temporary basis. Otor often occurs across aimag and sum boundaries, and thus represents a traditional component of Mongolian pastoral nomadism that may defy administrative boundaries and regulations. In June 2005, for instance, I came across two herders on otor who were moving flocks of sheep and goats eastward from eastern Zavkhan Aimag with a final destination of the Darkhan area. They claimed to be covering 40 kilometers a day (see figure 2.4).

Nonemergency otor may be motivated by the desire to fatten up livestock on distant pastures that offer better feed than the locally available pastures. Emergency otor, however, is a response to drought or other factors that pose immediate threats to herds. Going on otor requires good organization and the willingness to split the ayil into two groups—one that stays with the remaining, nonmigrating herd animals, and one that goes on the long-distance migration with the designated animals. But going on otor is the exception, not the rule.

Figure 2.4 Zavkhan Aimag herders on otor, 2005.

When one looks at pasture migration patterns, one sees that herding households in the same ayil generally migrate en masse from one seasonal pasture to the next. One important reason for moving camp in a group, according to Natasha Fijn's research, is that herders believe that wolves would descend upon a small herd if a single household were to remain behind the main migrating group.[43]

With all the regional diversity in Mongolia, it is, upon reflection, amazing that a tradition-based pastoral nomadic system of herding animals continues to be a successful venture throughout much of the country. The products of Mongolia's livestock industry constitute roughly 30 percent of the nation's gross national product, and about 40 percent of Mongolia's workforce is involved in animal husbandry. The processing and marketing of the dairy, hide, wool, cashmere, and other products of Mongolia's herds could certainly benefit from major increases in Mongolian governmental and international investment. This point is driven home when one reads that about one-third of the milk products of Mongolia's herds spoil on the way to processing facilities.[44] Construction of local milk processing enterprises would also allow herders to sell locally rather than lose value through middlemen. Similarly, it is astonishing that about 80 percent of Mongolian raw cashmere ends up being exported to China instead of being processed in-country.[45] It is the country's infrastructure—roads, factories for processing dairy, cashmere, wool products, and so on—that is at fault, not the products of the herders. Progress is being made, however, toward moving Mongolian livestock products into the world market. The plan to register all Mongolian livestock with the World Organization for Animal Health is the first step toward the eventual goal of selling meat on the world market. By 2012, 15–20 percent of Mongolia's livestock is expected to be registered; by 2015, 40 percent; and by 2021, 80 percent. Registration will not only contribute to food safety within Mongolia, but will also allow herders to benefit from inclusion in the international sale of meat.[46]

As one researcher has recently pointed out, Mongolia's "harsh climatic conditions are not a constraint; they are the reason for the extensive, mobile, animal production based exclusively on natural pasture, that has proved sustainable over many centuries."[47] Sustainability and modernization may prove complementary in Mongolia's pastoral nomadic economy, if economists, politicians, and other Mongolian citizens accept the premise of pastoral nomadism's historically proven adaptability and flexibility. The following chapter will seek to provide a historical background on Mongolia's nomads, their conceptualization of rights over land, and their relationship with secular and religious authorities in determining land use from the thirteenth century to the early twentieth century.

CHAPTER 3

EARLY HISTORY

THE THIRTEENTH CENTURY

Rather than totally free, unrestricted access to pastureland, Mongolia's pastoral nomadic herders through history have often had to contend with some form of authority in determining pasture access and land use. Even in the stateless conditions of the pre-Mongol Empire period, it is likely that tribal and clan authority had a large role in determining households' land use patterns. One Mongolian scholar, Bat-Ochir Bold, has offered the generalization that even without strict borders between pasture zones, the different Mongolian and Turkic tribes in the era before Chinggis Khan "followed the ancient tradition, or the optimal method, of pasture land use" and lived for the most part in harmony.[1] Of course, such a generalization is difficult to prove or disprove, and no definition of "ancient tradition" is offered.

Thirteenth-century sources are not generous in the information they provide about land use per se. When we look at those few passages that mention authority over pastureland in the *Secret History of the Mongols* (1228), the earliest written history of the Mongols by an anonymous Mongolian author, we may get some sense of how rights of access to pastureland could be allocated by a supratribal leader. In *Secret History* §219, Chinggis Khan rewarded Sorqan Šira, who several years earlier had helped shield the young Temüjin from the hostile Tayiči'ut. Chinggis granted Sorqan Šira "free use of [the] grazing grounds" that he requested: "Činggis Qa'an said, 'Settle on the territory of the Merkit on the Selengge and, indeed, have free use of its grazing grounds.' "[2] Thus, pastureland that had once belonged to the Merkit tribe, a tribe defeated in battle by Chinggis Khan, was given to reward past service and to encourage ongoing loyalty to the Khan.

In *Secret History* §279, Chinggis Khan's successor, Ögödei Khan, in a decree full of largesse, ordered that grazing lands and water sources (*nuntuq usu*) be given "to the people (*ulus irgen*)."[3] And, while enumerating his four good deeds in §281, Ögödei lists the third of the four as his having "had wells dug in places without water and [having] had the water brought forth, thus providing the people with water and grass."[4] The clear implication here is that a responsible Khan will make sure that pastureland and water will at least be sufficient, if not abundant, for his own people. Yet, we still do not have a picture of how access to pasturelands was determined among households, clans, and tribes. The *Secret History of the Mongols* has many more passages that refer to the rewarding of loyal followers with captive peoples, not lands. In early thirteenth-century Mongolia, land was not conceptualized as an asset to be carefully divided up and parceled out.

It is also very important to stress that authority over pastureland, as outlined in the *Secret History*, does not equal ownership per se of land. Most Western historians nowadays avoid the terminology of feudalism when describing Mongolian society of the thirteenth and fourteenth centuries. The Soviet-era theory of "nomadic feudalism," first developed by B. Ia. Vladimirtsov in his 1934 work *Social Structure of the Mongols: Mongolian Nomadic Feudalism (Obshchestvennyi stroi mongolov: mongol'skii kochevoi feodalizm)*, was based on the core notion of nomadic land ownership, a notion unsupported by the sources.

The *Secret History* also has several passages relating to hunting practices, and some of these passages can offer insight into how control over territory was conceptualized in the era of Chinggis Khan and his sons. In *Secret History* §281, Ögödei's confessional list of his four faults includes his fencing in of hunting grounds to prevent wild game from straying into his brothers' territories. He confesses to "being greedy."[5] While reserving animals for large-scale hunts was a common practice among the Mongol khans, as Igor de Rachewiltz and other scholars have pointed out, putting up fences and pounded earth walls, as Ögödei describes his actions, apparently went one step too far. Fences and walls were foreign to Mongolian culture in the thirteenth century and to some degree they remain so today.

Claiming rights over land for hunting purposes in the thirteenth century could lead to great animosity and resentment, as §9 of the *Secret History of the Mongols* suggests: "As in their land the Qori Tumat had imposed bans on one another's sable, squirrel and wild game hunting grounds, and mutual relations were bad as a result, Qorilartai Mergen separated from the Qori Tumat and took the clan name Qorilar."[6] In this case, a dispute over hunting territory led to a split within a clan with a new clan emerging and relocating.

There is a tendency in the modern environmentalist literature on Mongolia to insert current environmental conservation values back into the thirteenth century. For instance, an article on Mongolia's Protected Areas credits Chinggis Khan with having "created hunting reserves to protect game species."[7] It is more likely that Chinggis Khan marked as "off limits" certain areas abundant with game in which he and his royal entourage wished to hunt. Chinggis Khan in *Secret History* §187, for instance, allowed two loyal servitors, Badai and Kišiliq, to keep whatever wild animals they killed while hunting in a battue (a hunt in which game is driven toward one area). This passage strongly implies that it was the Khan's prerogative to claim all such game. In order to inspire or reward loyalty, Chinggis Khan would generously authorize his followers to keep the harvest of the hunt. In §199 of the *Secret History*, Chinggis Khan directed Sübe'etei to limit his troops' hunting of wild animals, admonishing him to "hunt with moderation."[8]

Conservation of wild game through placing limits on hunting was motivated by the realization that over hunting would adversely affect the Khan's future hunts. Wild game, like urban loot, came under the Khan's jurisdiction. Thomas Allsen, in his masterful monograph on the royal hunt, has also written of "an inherent and permanent tension between the desire to amass triumphs over animals and the desire to preserve stocks for further displays of hunting prowess."[9] Conservation, thus, was a sort of accidental by-product of royal or Khanly greed.

Chinggis Khan displayed great anger—even at his own sons—when resources obtained from conquest were not properly handed over to him. For instance, in April 1221 when three of Chinggis's sons, Joči, Ča'adai, and Ögödei took the city of Gurganj (Urgench) in modern-day Turkmenistan, "they shared the people of the cities among all three of them but did not give Činggis Qa'an a share." (*Secret History* §260).[10] This resulted in furious admonitions from the Khan. Thus, we may detect clear parallels in how the Mongols viewed control over pastureland, control over wild game hunting territories, and control over conquered peoples and their economic and cultural assets.

The Mongols in China: Concepts of Land Tenure in the Thirteenth and Fourteenth Centuries

The Mongolian conquest of China spanned three generations of khans, from Chinggis (d. 1227) to his grandson Khubilai (d. 1294). Khubilai's reign (1260–1294) oversaw the consolidation of conquered territories within China proper and the establishment of the Yuan Dynasty in 1272. Thus, the Mongolian Yuan Dynasty ruled all of China from 1279 to 1368. Not surprisingly, the Mongols were exposed to long-established Chinese legal codes, modes of rulership, and particularly the weight of a dominant bureaucratic hierarchy. In response to Chinese traditions, the Mongolian khans crafted a system of ruling that blended their own customary laws, Chinese legal and administrative practices, and influences from the Central Asian peoples—Uighurs, Tibetans, and others—whom they employed to balance the potentially overwhelming Chinese presence.

When we examine Mongolian versus Chinese concepts of land ownership, usufruct, and land use, the contrast is quite stark: communal, nomadic land use for grazing livestock versus privately owned, sedentary lands for agricultural production. Clearly, the Chinese legal tradition mandated written contracts for any transaction involving land, such as sales, mortgages, rental agreements, and so on. Much of Yuan-era contract law seems to have been based on the legal practices of earlier Chinese dynasties, particularly the Song (960–1279).[11]

The Mongolian rulers of China made no attempt to interfere with the complex administrative details of land transactions on the local level; they were content to allow clerks (a poorly recompensed stratum at the low end of the Chinese bureaucratic ladder) to oversee such transactions, and when irregularities occurred, either community leaders or, if necessary, a county prefect would become involved.[12]

From the Chinese perspective, however, the Mongols were barbarians with little understanding of civilization. Expecting the worst of their overlords, Chinese historical sources record the tale (very likely apocryphal) that an advisor to Ögödei Khan (r. 1229–1241) dissuaded the Khan from annihilating the inhabitants of North China and turning all the region's grain fields into pastureland for livestock. The key argument of the advisor hinged upon taxation of agricultural land as a more reliable source of income than vast destruction.[13] Thus, the trend away from military rule to civilian rule began under Ögödei's rule.

Yet, North China did indeed suffer greatly in terms of loss of human life and destruction of crops in the conquest wars of the early- and mid-thirteenth century. By the time that Khubilai Khan ascended to the throne in 1260, however, the ruling Mongols seemed to understand that taxes levied on agriculturally productive lands were preferable to militarized rule and destructive forays. In fact, Yuan sources provide evidence that local officials by the 1280s were specifically authorized and expected to supervise and encourage agriculture in regions under their jurisdiction.[14] In 1264, for instance, both civil and military officials and troops were warned by imperial decree that if they were to inflict damage upon crops, mulberry and fruit trees, or cattle, the offenders should be assessed fines as recompense for the damage.[15] The Yuan Dynasty bureaucracy also included a so-called Grand Bureau of Agriculture that oversaw the printing and distribution of an agricultural handbook that was based on earlier such works. While not introducing new information per se, the Yuan handbook did serve to educate the Yuan officials who were mandated to promote the development of agriculture.[16]

Overall, South China witnessed a continuation of agricultural landholding patterns with no overt attempt at disruption by the Mongols. If one can fault the Mongols, it would be for excess vigor in increasing the number of registered taxpaying households in both North and South China in order to raise revenues for the Yuan treasury. Land taxes on South China's large estates, however, were not overly burdensome, since the Mongolian khans from Khubilai's reign on realized the important economic contribution of agricultural production.

When we search for a uniquely Mongolian conceptualization of landholding, North China provides such an example. In the North, the ruling Mongols created and lavishly supported semi-independent landholdings called *touxia*, a term often translated as "appanages" or "fiefs."[17] In contrast to the traditional Chinese bureaucratic impulse toward centralization, the Mongolian institution of touxia fragmented landholdings and control over touxia populations: imperial princes, military figures and others deemed worthy of rich rewards were often presented with touxia lands and population in recognition of their status or deeds. Conceptually, the touxia might be seen as "shares" of the conquered lands of North China. The touxia grantees not only could appoint their own officials, but they also could—to varying degrees in the course of the Yuan Dynasty—retain tax revenues for their private use instead of forwarding such revenues to the state treasury.

The fragmentation of political control among Mongolian princes in the Yuan period is highly significant in terms of suggesting the underlying nomadic counterpull against centralized (Chinese-style) authority. The decentralization reflected in the touxia system also was a harbinger of the intense conflicts among Mongolian tribal and clan confederations that occurred in the post-Empire period.

POST-EMPIRE MONGOLIA: SHIFTING ALLEGIANCES

With the Mongolian conquests of sedentary lands (China, Persia, and Russia) during the course of the thirteenth century, the history of the geographically separate Mongolian populations varied dramatically. The Mongols in the Iranian-Turkic cultural spheres largely converted to Islam and intermarried with local peoples during the fourteenth century. Because those Mongols who settled in Iran, Afghanistan, along the lower Volga, and elsewhere in the Islamic world did not return to the Mongolian homeland, they shall not be the focus of our attention. The Mongols who ruled China, however, did return to the steppes of Mongolia after the dynasty that they ruled (the Yuan Dynasty, 1271–1272) collapsed during a wave of rebellions in the 1360s.

During the course of the Yuan Dynasty, Mongolian rulers, soldiers, and officials within China faced a variety of cultural and social influences in the form of Confucian ritual, Chinese Buddhism, Tibetan Buddhism, and Daoism. As both Chinese and Western observers (including Marco Polo, the Venetian merchant who visited Khubilai Khan's court) have noted, the Mongols continued to adhere to their own shamanist rituals and beliefs. The Mongols are also renowned for having tolerated the religions of those whom they conquered, in part, at least, because such tolerance was explicitly predicated upon the conquered populations' willingness to remain peaceful and to pay tribute and taxes.

The Mongols' adherence to their own cultural norms led to an interesting outcome when, in 1368, the troops of the Chinese peasant rebel Zhu Yuanzhang approached the Yuan Dynasty capital of Dadu (Beijing): many Mongols fled north to their ancestral homeland, accommodating themselves as best they could to the nomadic way of life. A seventeenth-century Mongolian chronicle, the *Erdeni-yin tobči*, claimed that there were 400,000 Mongols in China proper at the end of the Yuan Dynasty, and that 60,000 fled northward while 340,000 were unable to escape.[18] Such a later Mongolian source would likely not have been able to provide exact numbers, yet it is clear that a very large exodus of Mongols occurred in 1368. They did not, however, return to an empty steppe. During the whole of the conquest period as well as the period of the Yuan Dynasty, Mongolian herding families had continued to practice their customary way of life. Because the Yuan rulers required a constant flow of military recruits for conquests and for maintaining garrisons, many Mongolian herding encampments during the thirteenth and fourteenth centuries were reduced to populations of women, children, and the elderly. The Yuan dynastic history (*Yuan shi*) records numerous instances of government subsidies of grain sent northward to alleviate the hardships suffered by Mongolian commoners in the homeland throughout much of the dynasty.[19]

When the Mongols who fled the Yuan capital in 1368 returned to Mongolia proper—specifically to the Kerülen River area—they clashed with local Mongolian chieftains and they undoubtedly overburdened the fragile local economy by their sheer numbers.[20] The merging of these two disparate populations of Mongols who had lived in such diverse environments—one sedentary and one nomadic—was a recipe for disaster. Political instability was to ensue for centuries, as would-be khans made various attempts (mostly unsuccessful) to reunite quarreling factions. The economic distress caused by the sudden appearance of such a great number of returnees led to food shortages and thus a recurring pattern of Mongolian raids upon northern Chinese border towns. The returnees had more extravagant tastes, higher expectations, and probably fewer survival skills, contributing even more to the instability of the post–Yuan Mongolian homeland.

The Ming Dynasty that ruled China after the collapse of the Yuan faced this Mongolian threat for most of its 276-year existence. While some Mongolian military ventures may have been calculated to restore the past glory of the former Yuan Dynasty, it seems more likely that many such episodes grew from the Mongols' desire to supplement the products of their pastoral economy with Chinese rice, grains, and tea, products to which the Mongolian elite had grown accustomed during their rule of China. There was even a steady, if not large-scale, flow of Mongols from the steppes into Ming China through the mid-fifteenth century. Those Mongols who fled the constant internecine fighting among Mongolian princes and periodic famines were welcomed by the Ming Dynasty, which willingly incorporated the new immigrants into the Ming army often as border troops employed to repel new waves of incursions by marauding Mongolian forces.[21]

The first decades of the fifteenth century saw large-scale Chinese military incursions deep into the Mongolian homeland. In Mongolia, the constant need for military manpower depleted labor needed for herding livestock; when battles occurred within Mongolia itself, pastures were destroyed and nomadic migration routes were disrupted. Even after such major Chinese invasions into Mongolia ended, a fragile and easily shattered balance of military power between the Ming Dynasty and the Mongols to the north ensued. The resulting social and political instability among the splintered Mongolian clans and tribes put control over pastureland in flux. It is not surprising that in this period of chaos, a Mongolian prince would turn to a new religion—Tibetan Buddhism—to bolster his prestige and to reunify the warring Mongolian factions.

BUDDHISM: ITS IMPACT UPON MONGOLIAN SOCIETY, POLITICS, ECONOMY

Altan Khan (1508–1582) had threatened the Ming Dynasty with several military raids, including one in 1550 in which his troops appeared below the walls of Beijing. In 1578, Altan Khan met with a Tibetan Buddhist patriarch in Köke-nuur and Altan's conversion to Tibetan Buddhism was accompanied by the formal recognition of him as a reincarnation of the Yuan Mongolian ruler Khubilai Khan. (In return, the Tibetan cleric had the title "Dalai Lama" bestowed upon him.)[22] While the conversion hinted at ambitions to reintegrate Tibet, Mongolia, and China (as they had been integrated as parts of the Yuan Empire of the thirteenth and fourteenth centuries), Altan Khan's greatest long-term influence was the reintroduction of Tibetan Buddhism among the Mongols themselves. Why is this important in a history of land use in Mongolian history? As we shall see, the Mongolian Buddhist establishment by the eighteenth century was immensely powerful and wealthy: both livestock ownership and control over the finer pastures would contribute to the monasteries' wealth.

The conversion of the Mongols to Tibetan Yellow Sect dGe-lugs-pa Buddhism proceeded from the top down, that is, from the princely elite to the commoners. Altan Khan, who could not read Tibetan or Chinese, realized the importance of sponsoring translations of Tibetan sutras into Mongolian (sometimes via Chinese translations).[23] After Altan's death in 1582, the Mongolian elite continued to request from the Ming court Mongolian translations of Tibetan Buddhist works, and the Ming obliged.

The allure of a religion steeped in a rich written tradition can be understood in part by Altan Khan's desire to cultivate his image as a legitimate ruler among his own subjects. Because he was a descendant of Chinggis Khan in the twenty-fifth generation, Altan Khan already had a firm basis for claiming the right to rule the Mongolian people. In addition, Tibetan Buddhism's historical connection to the thirteenth-century court of the Yuan rulers made it a compelling component in any attempt to revitalize Mongolian unity and imperial ambitions. Tibetan lamas had constituted an important presence at the court of Khubilai Khan and his successors in Yuan Dynasty times.

Altan Khan himself ordered the building of the first monastery in what is now Inner Mongolia (a part of China), while Abadai Khan of the Khalkha (Northern Mongols) commissioned the building of the Erdeni Juu monastery, constructed on the ruins of the Mongols' thirteenth-century capital, Khara Khorum. Not surprisingly, Chinese labor constructed the new monasteries; Mongolian herders would have had little expertise or experience in the construction of such permanent structures. Evidence of the Chinese role in construction is clear in a 1606 Chinese-language inscription on stone commemorating the building of a temple of Maitreya in Inner Mongolia. Significantly, the names of both the carpenter and the stone carver were Chinese, as one would expect.[24]

Some of the earliest Buddhist monasteries on the Mongolian landscape consisted simply of *gers* clustered together. The spread of Buddhism throughout Mongolia led eventually to a sort of proto-urbanism, with monasteries situated on trade routes and serving as market centers. As monasteries assumed a more permanent presence on the Mongolian landscape, the buildings themselves incorporated more Tibetan architectural detail. This paralleled a growing Mongolian interest in other aspects of Tibetan civilization—medicine, art, astronomy, and so on.[25] Tibetan culture, and more specifically Tibetan Buddhism, offered the Mongolian princes and their retinues an alternative to Chinese cultural mores. The new orientation toward Tibet did not seem to alarm the neighboring Ming Dynasty rulers, but, as we shall see, the later Manchu rulers of the Qing Dynasty (1644–1911) acted swiftly to disarm the potentially dangerous coalescence of secular and religious power.

As Buddhism spread throughout the Mongolian realms, shamanist rituals and objects of worship such as *ongghod* (figures made of felt) were suppressed and often destroyed. The story of how the Lamaist gods replaced shamanist idols, of how Buddhist prayers replaced shamanist songs, and of how lamas took over the healing functions of shamans has been recounted in detail by the German scholar Walther Heissig in his classic volume *The Religions of Mongolia*.[26]

Mongolian nomads were cajoled and threatened into relinquishing their shamanist idols, as Lamaist missionary activity spread and intensified among the Mongols in the following centuries. As Heissig has pointed out, the Mongolian model of conversion followed the previous Tibetan model: the gods of the preexisting Tibetan Bon religion had been incorporated into Tibetan Buddhism, and so Mongolian shamanist rituals and gods were subsumed into the mélange of shamanist influenced Lamaism. Yet, many rituals of Mongolian folk religion, such as prayers and offerings to Chinggis Khan, to mountain deities, and to the *Chaghaan Ebügen* ("White Old Man") survived the Buddhist transformation, largely because such family-based practices did not require the participation of a shaman or a Buddhist lama.[27]

The economic impact of the spread of monasteries came to be felt by herders particularly from the seventeenth century forward. On the positive side, one can see the monasteries as mini-urban locales that functioned as periodic market centers, promoting the exchange of goods and services and often offering space for festivities. It is safe to say that Buddhist monasteries initiated city-building activities on the Mongolian landscape.

On the negative side, Mongolia's monasteries have been portrayed as amassing enormous wealth in land and livestock and as controlling a workforce sometimes termed *shabi* (a Mongolian word that may be translated as "disciples" or "serfs") or lay disciples gifted by Mongolian nobles to a temple or monastery. On the issue of land ownership by Mongolian Buddhist monasteries, scholars remain divided. The classic Marxist argument was put forward long ago by the Mongolian historian M. Sanjdorj in his book, *Manchu Chinese Colonial Rule in Northern Mongolia*:

> In the sixteenth and seventeenth centuries... Mongolia was a nation where feudal relationships were strong and highly developed. The land—the most important tool for production—was the property of the feudal classes. As a result of the character of the nomadic economy

> under conditions where the commodity-money relationship was very loosely developed, feudal land ownership manifests itself in the full right to choose the nomadic camps and divide the pasture land.[28]

A contrasting point of view may be seen in Robert Miller's work on the Buddhist monastic economy in Mongolia. Miller's 1952 interview with a high-ranking Buddhist reincarnation, the Dilowa Khutukhtu, included this interesting observation on the relationship between a monastery and the land:

> Even the land on which the monastery stood was not the property of the monastery; only the services of the families attached to the monastery, the buildings of the monastery itself, and the cattle (horses, sheep, goats, camels) given to the monastery were its own.[29]

Miller, however, did not view the Dilowa Khutukhtu's portrayal of the monastery's relationship to the land as typical; Miller, whose work focused far more on Inner (or southern) Mongolia than on northern (or Outer) Mongolia, described monastic endowments that typically consisted of land as well as livestock herds. Nomads often donated livestock to monasteries, and since land donations would have been insufficient to pasture the donated herds, these herds would be cared for by herding families who then returned a set percentage of the dairy and wool products and newborn animals. When devout laymen donated lands to monasteries, those monasteries would often rent out the lands, receiving rents in either cash, grain (primarily in Inner Mongolia), or herd animals.[30]

Whether pastureland and livestock both fell into the same category of "ownership" per se may seem like a semantic or moot question, since ultimately one might agree with Sanjdorj's statement that "[o]wnership of livestock, which was an important asset in the livelihood of the herdsmen, became, together with the land, the economic basis for political rule."[31] In other words, whether or not Mongolian Buddhist monasteries owned the lands upon which their herds pastured, the monasteries did acquire enormous economic and political power by virtue of this dual handle of authority.

The positive versus negative effects of the monasteries' power have been much debated in the context of Qing Dynasty control over

Mongolia, a period when the arrival of Chinese merchants, farmers, and would-be colonizers strongly pushed the equation to the negative side. Nonetheless, the basic elements of this debate pertain to the pre-Qing period as well. George Murphy succinctly presented both sides of the argument in an article in which he was attempting to balance what he saw as Robert Miller's view of the monastery as "parasitic."[32] Miller's view, supported by many other scholars, both Mongolian and non-Mongolian, is that the monastery system was defined by bribery, corruption, and inordinate spending on projects that benefitted only the monastery, not Mongolian society as a whole. With so much of the male population taking monastic oaths, the Mongolian population as a whole was demographically stagnant or even on the slide toward extinction. Of course, other factors, particularly diseases, contributed to the demographic crisis that was brewing over the centuries. Miller and other scholars also have portrayed the monastery system as siphoning off the potential for intellectual and social advances.

Murphy, however, points out the positive impact of the monasteries' role in Mongolian history, particularly their ability to act as "primitive banks" and their facilitation of commerce by underwriting caravans and providing accommodations for merchants and travelers in general. Murphy finds evidence of monasteries promoting agricultural uses of the land—of course, with their own profit in mind. In terms of intellectual life, monasteries did provide avenues to literacy (in Mongolian and in Tibetan), and book learning in certain nonreligious fields such as law and medicine.

Certainly a number of young lamas who had received religious instruction in the Tibetan language returned to lay life and took with them a basic literacy—using Tibetan script to write Mongolian. Statistics on literacy and even on the numbers of lamas in pre–twentieth-century Mongolia are nonexistent. In 1918, however, it was calculated that 45 percent of Outer Mongolia's male population had at some point in their lives been enrolled as lamas; thus, it is not surprising that functional literacy in the Mongolian language in Tibetan script was fairly widespread.[33]

In Mongolia's temples and monasteries, lamas would also offer instruction to the children of nearby families. Children of herder families might thus have some opportunity to obtain a functional level of literacy through temple education, while children of Mongolian princes in the period of Qing hegemony (late seventeenth to early twentieth centuries) might be educated at *yamen* (Qing government posts). In order to avoid clerical service in the Qing bureaucracy, some literate Mongols sought registration as illiterate, thus making real estimates of literacy even more challenging in the pre-1911 period of Mongolian history.[34]

In addition to the Buddhist canon, many secular works such as astrological books, dictionaries, and historical chronicles were published in Mongolian from the late sixteenth century on, suggesting a growing readership.[35] Walther Heissig, who traveled extensively throughout Inner Mongolia in the 1930s and 1940s, documented the immense variety and number of manuscripts and books that could be found in the gers of ordinary herdsmen.[36] Thus, in terms of literacy as well as other aspects of Mongolian society in the sixteenth through nineteenth centuries, the introduction of Buddhism and the spread of monasteries throughout the landscape had complex and long-lasting repercussions.

MONASTERIES, PRINCES, AND PASTURELANDS IN THE QING PERIOD

The remainder of this chapter will argue that in spite of the immense growth of centralized authority in the period when the Manchu Qing Dynasty exerted control over "Outer Mongolia" (late seventeenth century through 1911), communal and localized landholding patterns persisted. While Tibet's influence was apparent with the introduction of Buddhism in the late sixteenth century, China's impact upon Mongolia from the late seventeenth century onward eclipsed all other foreign influences. To understand the growth of Manchu-Chinese hegemony over the Mongols, a brief review of early Qing expansionism is necessary.

First, one must understand the geographic designations. The nomenclature of "Inner Mongolia" and "Outer Mongolia" is Chinese nomenclature, not Mongolian. The terms for the two Mongolias derived from the Chinese perspective on geography: "Inner" was closer to China proper, and "Outer" was far distant. Most Mongols in the country of Mongolia today would find the term "Outer Mongolia" offensive, since its use implies Chinese sovereignty.

The crescent of territory traditionally referred to as "Inner Mongolia" has changed over history mainly as a result of military invasions and political machinations. Thus, as the Manchu conquerors from the far northeast of China proper planned their campaigns to take Beijing and North China in the early- to mid-seventeenth century, they sought to neutralize the potentially hostile Mongolian tribes of Inner Mongolia. Letters from the Manchus to the Mongols of Inner Mongolia reflected a strategy of appealing to cultural solidarity (and the Mongols and Manchus did have much in common including a written script and aspects of a nomadic way of life) as well as rather direct threats should the Mongols not aid the Manchus. In a 1640 imperial missive from the Manchu ruler Hong Taiji to those Mongols who were living under Chinese jurisdiction, tents, cattle, and servants were promised to any Mongols who went over to the Manchu side.[37] Assuring that the Mongols would not form a hostile flank, the Manchus could then proceed with their invasion of North China, taking Beijing in 1644 after decades of preparatory battles to the northeast.

The Qing Dynasty, having consolidated power internally within China through the seventeenth century, also faced rebellion along another frontier—modern-day Xinjiang where the Zunghars, a western Mongolian confederation, resisted Qing control. When the Zunghars attacked the Khalkha Mongols of Outer Mongolia in 1688, Qing armies intervened on behalf of the Khalkhas and defeated the Zunghars. As a result, in 1691, the Qing convened an assembly of Khalkha (i.e., Outer Mongolian) princes at Dolonnuur in Inner Mongolia and orchestrated the official submission of Outer Mongolia to Manchu Qing hegemony. Not surprisingly, the so-called Dolonnuur Convention or Assembly has been judged by some modern-day Mongolian historians to be a low point in Mongolian history, and the princes who submitted to Qing control have even been condemned as traitors. As the British historian Charles Bawden has pointed out, however, the Khalkha princes had little choice in the matter and scant ability to foresee what their actions would bring.[38]

Meanwhile, the Manchus were also instituting enormous changes in the geographic administration of the territories of both Inner Mongolia and Outer Mongolia.

These changes stemmed from a military-demographic system called by the Manchus the banner system. Instituted in order to fill the ranks of its conquest units, the banner system initially was restricted to the Manchu people, but then, as Chinese and Mongolian populations were conquered and incorporated into Manchu-controlled territories, Chinese banners and Mongolian banners resulted. In Inner Mongolia, 49 banners had been constituted by 1670, and banner populations were required to supply soldiers to the Manchus.[39] The 1691 submission to the Manchu Qing Dynasty by the Outer Mongolian Khalkha princes led to the instituting of 34 banners in Outer Mongolia, a number that grew to 86 by 1759. While Christopher Atwood has written that the banner system "did not change traditional Mongolian social structure,"[40] most other scholars have argued that the imposition of these new territorial units (banners referred to both the people and the territory) led to restrictions upon herders' mobility and choice of pastureland.

As Peter Perdue has suggested, the Manchu rulers used the banners to tighten up control over the subdued Khalkhas. The territory of each newly instituted Khalkha banner was to be strictly demarcated and movements between pasturelands were to be regulated by the Manchu administration. To placate the Khalkha princes, they were allowed to keep their titles as khans and they were rewarded for loyalty by being given new titles and ranks by the Manchu rulers. Some Khalkha khans rejected the notion that a higher authority would have control over seasonal pastoral nomadic migrations.[41] The Inner Mongolian historian Sechin Jagchid has agreed with this assessment of the effects of the Qing banner system in Mongolia: "Strict boundaries were placed upon each banner, and the *jasagh* [banner head] as well as the people could not transgress these boundaries in herding and hunting" unless given specific permission by banner or imperial Qing authorities.[42]

Another well-known Mongolian historian, O. Pürev, has provided a chronology of Manchu encroachment: "In 1773, the Manchu administration legally established pasture boundaries for all the Mongolian *aimags* and banners. After establishing them in this way, in 1781 they set banner boundaries within Northern Mongolia (*Ar Mongol*) and they erected boundary *ovoo* as markers on the ground."[43] Yet even with limitations upon movement from one banner to another and defined pasture usage within each banner territory, land was apparently still conceptualized as communally held, not as privately owned. Even though Sechin Jagchid employs the terminology of feudalism, as in "a quasi-feudalistic administrative system" and "feudal-like system," he does not see any evidence of land ownership in the period of Manchu hegemony. In Jagchid's own words, "Although the people were confined to a particular area, and the people and the land within a banner were under the administration of the hereditary post of *jasagh*, the land was still considered to be communal; it was not the private property of the *jasagh* himself."[44]

Yet there was a clear trend from the eighteenth century to the early twentieth century toward much of the best pastureland being removed from general public use by both secular and religious authorities, that is, by the Mongolian princes and by the Mongolian Buddhist monasteries. The Mongolian historian Bat-Ochir Bold has written that the higher-ranking princes and the more influential Buddhist dignitaries were able to mark off the finest pasturelands for their own use, thus reducing unclaimed communal pastureland to about 40 percent of the total land viable for grazing herds.[45] The actual manner of demarcating pastures in earlier centuries involved the use of cairns of rocks (ovoo) that could serve as border markers but also served in other contexts as spiritual symbols and as the loci of sacrifices.[46] In his research on banner boundaries in late Qing-era Inner Mongolia, Henry Serruys delved into documents that showed the people of one Inner Mongolian banner moving earthen landmarks in order to expropriate territory from a bordering banner.[47] One has the sense that it would not have been overly difficult to manipulate such pasture boundaries on the sly (see figure 3.1).

The Manchu-imposed banner system fragmented the clan and tribal loyalties of the Mongols, erasing horizontal ties of loyalty and replacing them with vertical ties to the Qing rulers in Beijing, thereby greatly reordering Mongolian society. The Manchus were determined to prevent the reemergence of a Mongolian nomadic confederation

Figure 3.1 Ovoo near Hag Lake, Zavkhan Aimag, 2005. Nowadays, most ovoo are situated at sites of natural beauty (mountains, lakes), atop mountain passes, or at Buddhist sacred sites.

that might challenge Qing imperial rule. Thus, neither Mongolian princes nor Mongolian commoners were permitted to graze their herds or hunt beyond the prescribed boundaries of their own respective banners' common pastures. Violations were subject to punishments. As of the mid-eighteenth century, the Manchus also insisted that the highest ranking incarnate lama among the Khalkha (Northern Mongols) not be Khalkha himself but rather from Tibetan ethnic background. This of course was a blatant attempt to separate sources of secular authority (Khalkha princely lineages) from sources of religious authority in the Buddhist hierarchy. Compliant Mongolian princes were also rewarded for their loyalties to Beijing through intermarriage with the daughters of the Manchu imperial line and through the continual bestowing of ranks and titles.

Ultimately, the Manchus never attempted to alter the basic foundations of traditional Mongolian nomadic practices. Considering that the Manchus governed a mostly agrarian empire with a socioeconomic system based on private land ownership, there is no doubt that the Qing rulers had a good grasp of the benefits of taxing and controlling a sedentary population. One might ask why the Manchus did not tinker more extensively with land use in the nomadic zones of their empire. The Qing Empire was powerful indeed, but the Manchu colonial rulers of Mongolia had good reason to refrain from radical experimentation in land use patterns. The Manchus themselves were an Inner Asian people with seminomadic roots and, in spite of ruling over an agrarian peasantry in China proper, they undoubtedly realized that it would not be worth the trouble or enormous resources that would be entailed to transplant an agriculturally based system of private landholding upon a traditional herding society.

There was also a military rationale for not tampering with traditional Mongolian society in Qing times. As Charles Bawden has noted: "The general line of Manchu policy was to keep Mongolia as a military reserve, and to effect this the successive emperors tried as far as possible to seal her [Mongolia] off from outside contacts, and to preserve the traditional pattern of nomadic pastoralism."[48] Of course, more or less indirectly, Manchu rule did have many negative consequences for Mongolia, particularly in the realm of Qing taxes, spiraling indebtedness to Chinese merchants, and resulting famine and vagrancy among Mongolian commoners.[49]

Even though the Qing Dynasty attempted to restrict and regulate Chinese merchants who went to Mongolia in ever-greater numbers in the eighteenth and nineteenth centuries, the constantly expanding operations of Chinese merchants plunged much of Mongolian society, from princes to commoners, into deep indebtedness. It is quite common in the modern Mongolian historical narrative to come across references to "the debt to the greedy Chinese merchants" and to livestock "appraised by the Chinese at prices below local prices."

The resentment toward Chinese merchants that was felt by Mongols born in late Qing times continues to color autobiographical and historical works by Mongolian authors.[50]

The impact of the Chinese presence in Mongolia was felt on the land itself. The sources speak of Chinese merchants and farmers illegally colonizing lands in various parts of Mongolia. For instance, an 1808 document complains that a banner prince (in modern-day Bulgan Aimag) had "allowed the Chinese people, who were permitted to stay here for a limited time, to build up homes and enclosures and allowed them to settle down."[51] This clearly contravened Qing regulations that prohibited Chinese merchants from settling outside Urga's trade district. An 1897 document pertaining to Dornod and Khentii Aimags includes the complaint that "some Chinese were allowed by the officials to build houses, breed livestock and farm vegetables in the banner [Sečen Qan Aimag]. Moreover, some of the Chinese were allowed to stay in the banner, even though they had no license."[52]

A Russian observer traveling through northwest Mongolia in 1910 noted the extent of Chinese conversion of pastureland to cropland. I. M. Morozov, one member of the 1910 Moscow Trade Expedition, en route from Kyakhta to Urga, described a well-kept Chinese farmstead. The Chinese farmer had come from Datong and had settled in Mongolia 14 years earlier, in 1896. Morozov wrote that the Chinese farmer "received land for the cultivation of grain from the Mongolian banner prince with the assistance of the Urga *amban* [representative of Manchu Qing authority]. The best lands in the valley turned out to be at the disposal of the Chinese; the sand dunes and bare hills were left for the Mongols. The colonization of this area by the Chinese is progressing quickly, and in the near future the Mongols will be completely displaced from here."[53]

Morozov also noted the potential for profit in farming in Mongolia: "The practice of agriculture in Mongolia at present [in 1910] results in great profits... Farming, with high market prices for flour in Mongolia, for a long time already has afforded a good profit, which is explained by the increased development of agriculture both among the Mongols and among the Chinese along the Selenge River and its tributaries in the region [spanning] Kyakhta—Urga."[54] Those Russian merchants residing in Urga greatly envied the Chinese success in practicing agriculture and marketing their harvests in Mongolia; the Russians had no treaty rights in Qing times allowing them to own or rent lands for the purpose of agriculture. Morozov described with great admiration the intricate systems of irrigation that he observed in Chinese gardens in Urga's Maimaicheng ("trade town"),[55] and he was amazed at the successful 1910 wheat harvest around Uliastai in western Mongolia, a feat accomplished in "adverse conditions of soil and wheat" in his words.[56]

Some Mongols did indeed try their hand at agriculture with mixed results, but this was far more common in Inner Mongolia than in Khalkha territory. Sechin Jagchid has described *Mongghol tariya* (a uniquely Mongolian form of farming) practiced with a very long-handled sickle, allowing the Mongol farmer to stand upright while cutting grass; seeds were then cast by hand and herds were kept away from the "sown" area.[57] As Jagchid notes, the Mongol farmer typically preferred not to use the Chinese-style short-handled sickle because of the aversion to stooping down and actually digging into the earth. Historical photos do, however, attest to the fact that Mongols planted and harvested grain near the western Khalkha towns of Uliastai and Khovd (Kobdo) in the early twentieth century.[58] For the most part, Mongolian farming did not rely on irrigation techniques, instead relying on nearby sources of water (rivers, springs). It was what might be termed a very passive approach to agriculture.

It has been estimated that by the late nineteenth century, roughly 150,000 or so acres were being farmed by Chinese farmers in the Selenge River valley in northernmost Khalkha territory.[59] The geographic spread of Chinese colonization was hindered only by the Russian border. The Qing Dynasty's tolerance of Chinese merchant and farming activity in Mongolia may very well have been intended to subvert and manipulate traditional Mongolian society through economic impact, which is different from using direct political or military means to control a society. There are obvious parallels in Sino-Mongolian economic relations between the Qing era and today: nowadays, Chinese investors buy large stakes in Mongolia's mineral and energy resources.

Just as land use responded to demands by exterior authority, livestock herds have adapted to pressures from exterior sources too.

Mongolia's pastoral nomadic herders have through history adapted their herd composition on the basis of fluctuating demands for some species over others. In chapter 1, I made reference to the concept of "ideal herd composition" in modern-day Mongolia. It is well nigh impossible to apply this concept to earlier centuries, since statistics for herd composition in the period from the thirteenth century to the early twentieth century are nonexistent. Yet, we may surmise that in the thirteenth-century Mongolian conquest period, horse and camel numbers were ascendant. With an estimated five mounts for every Mongol soldier and with horses and camels in demand to haul invasion armies' supplies, nomadic families from whom the livestock was requisitioned would clearly have had to adjust their livestock ratios.[60] Similarly, as Mongolia developed important trade centers along the so-called Tea Road between China and Russia in Qing times, camels were the beast of choice to convey bricks of tea and other goods from China to Mongolia and Russia.[61]

The loss of grazing lands for livestock as agriculture spread was the major trend in late Qing times. Even though Qing policies toward Mongolia never called for pastureland privatization, clearly the conversion of more and more land from livestock-sustaining pastureland to agricultural cropland, if continued long enough, would have irrevocably altered Mongolian herders' access to land, migration routes, and water sources. Qing policies undoubtedly resulted in the growing impoverishment of Mongolia's herders and the loss of access to land, often the best land, to Chinese colonizers. Yet, the 1911 demise of the Qing Dynasty and the resulting curtailment of Chinese economic activity in Mongolia in the following decades saved the traditional Mongolian pastoral nomadic way of life.

CHAPTER 4

THE SOCIALIST ERA

PRELUDE TO SOCIALISM: 1902–1921

Manchu policy toward the Mongols of Northern (Outer) Mongolia was never benign; rather, it was designed to balance the triumvirate of interests—Manchu, Chinese, and Mongol—with the least weight given to the Mongols. In 1902, Manchu policy took a sharp turn toward colonization and tighter administrative control of Inner Mongolia. The so-called New Administration as it was implemented in Inner Mongolia was designed to restructure the region into a Chinese province. In 1906, the policy was extended to Northern (Outer Mongolia) with the ultimate goal of administratively subsuming this culturally, historically, and environmentally distinct area.[1] Han Chinese were to be encouraged to move not only into Inner Mongolia, but also into Northern Mongolia to cultivate the land for agriculture, that is, to colonize the land.

Why the Manchus would revise their previous "hands-off" policies vis-à-vis pastoral nomadism and land use in Mongolia and instead push an agricultural development policy is an intriguing question. Being subjected to colonial bullying by Western powers and by the Japanese, the Manchu Qing Dynasty in the early 1900s had its back against the wall in terms of international pressures. In the realm of economics, the population of Qing China had roughly tripled between 1650 and 1850. By 1900, the impoverished, land-hungry subjects of the Qing were desperate for land to farm—and they were increasingly restive, with anti-Qing foment occurring on a regular basis. Redefining Mongolia administratively to consolidate Qing control in the face of Tsarist Russian economic interests and possible expansionism in Central Asia made eminent sense, as did the use of Mongolia as a safety valve for restless, land-poor Chinese.

Would such a grand scheme to introduce extensive and intensive farming into Northern Mongolia have been viable? First, it is important to acknowledge the fact that parts of Northern Mongolia have historically always supported limited agricultural pursuits. The discussion of agriculture in chapter 6 will elaborate upon Mongolian nomadic-style farming that usually entailed sowing seeds for millet or barley crops, moving away with the herds, and then returning later to harvest these crops. In contrast, Chinese farming techniques in Mongolia required transforming pasturelands into irrigated, intensively cultivated farm fields. Chinese farmers had made enormous inroads into Northern Mongolia. During the course of the nineteenth century, these farmers easily evaded the largely unenforced Qing legal restrictions meant to prevent Chinese settlers from seizing lands from Mongolian herders. Along parallel lines, Qing restrictions on Chinese commercial operations in Mongolia through Qing-issued permits to traders (a system that dated from 1720) were also ignored by Chinese traders who easily manipulated Mongolian herders into buying on credit. By 1911, the population of Northern Mongolia owed Chinese traders 15 million *tael* (units of silver).[2] The events of 1911 that led to a brief interlude of Mongolian independence also precipitated the mass exodus of Chinese farmers and a downward spiral in the number of acres devoted to agriculture in Mongolia.

Between July and December 1911, a group of Mongolian princes decided to push for Mongolian independence from the then teetering Qing Dynasty. While Russian assistance was sought, the actual secession from the Qing was engineered by the Mongols themselves. A theocracy or theocratic monarchy with the Eighth Jebtsundamba Khutughtu (1870–1924) as a religious/secular leader was declared on December 29, 1911. Independence, however, proved difficult to defend, and the three-power Treaty of Kyakhta (China, Russia, and Outer Mongolia) in 1915 demoted Mongolia's independence to autonomy under Chinese national sovereignty. Russia, embroiled in World War I, pressured Mongolia to participate and to agree to terms that benefited only Russia and China. In 1919, a Chinese warlord, Xu Shuzheng (1880–1925), nominally an appointee of the Chinese Republic, took over the administration of Mongolia, disarmed the Mongolian military, and imprisoned the Jebtsundamba Khutughtu, thereby transforming autonomy into subjugation.[3]

THE 1921 REVOLUTION AND ITS AFTERMATH

The political chaos of the period 1911–1921 ended only with the consolidation of power by the Soviet Red Army and its Mongolian supporters in the 1921 Revolution. Nationalism and anti-Chinese sentiment combined to produce a Mongolian revolutionary cohort, many of whom were well versed in the Russian language and willing to seek support from, and follow the policy directives of, the Soviet Union. Russianized Buriat intellectuals were also quite influential in the formation of a Mongolian revolutionary movement. The invasion and brief occupation of Urga by the White Russian commander Baron Roman F. Ungern-Sternberg and his anti-Bolshevik troops in early 1921 gave the Red Army in conjunction with Mongolian troops the pretext in summer 1921 to enter Urga, defeat the Baron's army, and establish a new regime or "Provisional Government" with the Jebtsundamba Khutughtu proclaimed a "constitutional monarch."[4] In spite of the initial promise that Red Army troops would be withdrawn once the Baron's anti-Bolshevik forces were defeated, Soviet troops remained in Mongolia for years. Once the Mongolian People's Republic was declared in 1924, Soviet civilian advisers flooded in, and the Soviet military presence was downgraded. The position of the Jebtsundamba Khutughtu as head of state ended with his death in 1924; by 1929, the Mongolian People's (Revolutionary) Party felt sufficiently empowered to decree that there would be no future reincarnation of the Jebtsundamba Khutughtu or of any other Buddhist ecclesiasts.

During the decade of the 1920s, Mongolian politics were in upheaval: the country's top leaders faced the difficult choice of either subordinating themselves to their Soviet advisors or of being purged from power. The economy, however, generally prospered. Livestock increased from 9.6 million in 1918 to 13.8 million in 1924; by 1930, the figure stood at 24 million.[5] The vacillating livestock figures in the decades of the 1920s-1930s directly reflected the extreme policy shifts of the new Soviet-directed regime. In 1929, the Soviet Union adopted the policy of forcing Soviet peasant farmers into collectives in order to dismantle "capitalist" property ownership. Mongolia followed suit with the first abortive attempt at livestock collectivization beginning in 1930. This met great resistance from herders and led to a steep decline in livestock numbers from 24 million in 1930 to 16.2 million by 1932.

From 1932 to 1936, collectivization was reversed and herds were reprivatized under the "New Turn Policy," allowing herd numbers to rebound to about 22.6 million by 1935.[6] Mongolia's so-called New Turn Policy was Stalin's response to the catastrophic economic and social toll that the 1930–1932 Soviet-modeled attempt at collectivization had caused. Mongolia's herders thus were treated differently from the Soviet Union's own population. To avoid collapse, Mongolia was granted a brief respite, unlike Soviet peasant farmers and Soviet Kazakh nomadic herders for whom unabated forced collectivization led to millions of deaths through the 1930s.

Religion in Mongolia faced severe persecution, mirroring the atheism campaigns of the USSR. In the Soviet Union itself, the forced collectivization of agriculture was closely tied to the complete elimination of religion and the secularization of society. The Russian Orthodox Church still held huge tracts of land in 1928, and the expropriation of Church lands was simply one aspect of Soviet collectivization of privately owned properties in the 1920s and 1930s.[7] Mongolia's Buddhist establishment, which controlled vast resources including pastureland and herds, faced persecution in the 1920s, a brief respite in the early 1930s, and finally, from 1932 onward, dire campaigns to destroy the monasteries themselves, to execute or imprison monks, and to redistribute monastic wealth and herds through collectivization. As Christopher Atwood has noted, Stalin saw Mongolia's Buddhist establishment as "a state within a state" that could pose an ongoing threat to the evolving socialist nation.[8] The purges of intellectuals, party members, military officers, and commoners, including herders, continued from 1933 until 1953, resulting, according to one Mongolian historian, in the deaths of an estimated 36,000 people who had been deemed anti-regime in one way or another.[9] While this number may seem small in comparison to the millions of Soviet citizens who died as a direct result of Stalin's purges, one must remember that the population of Mongolia in 1956, when the worst of the purges had subsided, stood at only 845,500.[10]

World War II offered herders a reprieve from collectivization, as Soviet policy gave priority to social and economic stability. Stability in Mongolia was important both for economic purposes—enabling imports of meat, butter, and other livestock products into the USSR—and for military purposes as Japan's expansionist probes into the heart of Eurasia posed threats to this flank of the USSR. Yet, the war years did inflict great demands upon Mongolia's herders, even though collectivization had been suspended for the time being. J. Sambuu, who served as MPR ambassador to the USSR in the wartime years, listed as "a gift from the Mongolian people" a hundred or so railroad cars filled with "15,000 military short sheepskin *deels* [traditional Mongolian garment]; felt boots; warm gloves; warm jackets; wolf, fox, and goat skins to make into warm clothing for airmen; horse hide belts and bags; sheep, cow and antelope meat; fat, butter, and foreign spirits."[11] An official MPR history of this period glosses over the privations suffered by the Mongolian herding households during the war years by accentuating the "voluntary" nature of their contributions to the Soviet Union:

> The herdsmen (*ard*) [*arad*] donated to the Red Army over 32,000 of their best horses and also sold half a million horses for the front at the state fixed price. The working people voluntarily donated to the Aid Fund valuables and private savings amounting to millions of *tögrög*. During the war years, eight special trainloads of gifts such as foodstuffs and warm clothes were sent to the Soviet troops from the MPR.[12]

The Soviet Union did not authorize renewed collectivization of herding again until the late 1950s. Yet, Mongolia's herders felt new pressures from their government by 1949, when they were required to increase livestock numbers by 50 percent a year and to fulfill state quotas of meat, wool, and milk as set by the five-year plans, which Mongolia had instituted based on the Soviet centralized economic model. There was little or no leeway for unpredictable events like *zud* in this new economic order.[13] It is worth stressing that zud was the wild card in the traditional economy of the Mongols and remained so during the socialist period as it does in the postsocialist era. Actual recollectivization was jump-started in 1955 when new regulations detailed how herders and livestock would be integrated into collectives (*negdels*). For instance, the ownership of most livestock would pass into state hands, but a fixed, and very small, number of herd animals could remain in private hands: 100 head in the Gobi regions and 75 head in the Khangai zone, for example.

Those herders who attempted to resist collectivization by holding on to their private herds were subject to much higher economic quotas than herders who joined collectives; often the state would requisition as much as twice as much milk per privately held cow or twice as much wool per privately held sheep as compared to quotas set for the produce of collectives' herd animals.[14] The use of economic coercion to push herders into collectives was, of course, a softer strategy than had been applied in the first collectivization attempt decades earlier, and it proved effective. Contrary to what one might expect, herders in the 1950s collectivization drive did not select the best animals to keep as privately owned, but rather kept the oldest and least productive ones. Apparently herders assumed that they would lose these animals in the next round of ongoing collectivization; in the meantime, they could slaughter them for meat.[15]

The process of collectivization, which had been attempted too rapidly in 1929, this time was pronounced successful by late 1959.[16] Mongolia had apparently fulfilled the goal set by Lenin in 1921: bypassing capitalism and thus progressing directly from feudalism toward socialism in the Marxist linear vision of human economic development. The much-heralded meeting of Lenin with the Mongolian revolutionary leader Sukhbaatar (1893–1923) and other members of a Mongolian delegation in 1921 in Moscow was the theme of several Mongolian paintings that were frequently reproduced in the popular press in the socialist era, though whether the two leaders ever did meet has been open to much speculation. The peculiar concept of bypassing capitalism was also popularized in Mongolian art, as in the famed mural by the Mongolian artist Dagdangiin Amgalan (b. 1933) that depicts a horse and a rider leaping over capitalism in a glorious manner.

Achieving socialism in a pastoral nomadic society called for creative interpretations of Marxist doctrine. In Mongolia's case, this meant that herds were collectivized, but the land was not. A closer look at the negdel system from 1959 to the early 1990s will clarify the relationship between the socioeconomic unit of the negdel and the land itself.

THE NEGDEL SYSTEM

"Generally, since herding is a unique profession, I believe that the 'book' about the herds should be 'written' in the pasture."[17]

"For economists and scholars, the pastures should be the real laboratory for creating new things."[18]

Both of these statements were written by Ts. Namkhainyambuu, the herdsman-author of an autobiography that delves deeply into the challenges of herding in socialist Mongolia under the negdel system. Both statements point to the gap that, in Namkhainyambuu's own words, existed between "life and theory" in socialist Mongolia. The negdel system was essentially an application of Marxist theories, theories that had been refashioned or twisted to fit widely varying circumstances in widely different societies—Russia, China, and, in this case, Mongolia. The theory of "nomadic feudalism," an attempt to retrofit class relations to a nomadic society, originated with the Soviet historian B. Vladimirtsov, who was a respected linguist and scholar of the Mongolian language. Mongolia's historians had to adhere to this theory of class relations in their writings in the socialist era.

Since Mongolia's social strata did not lend themselves easily to the Marxist-Leninist conflict scenario of feudal exploiters versus the exploited, Mongolia's secular princes and higher lamas were deemed the former, whereas herders were portrayed as "arad" or commoners. In other words, the herders were a substitute proletariat, since no urban working class existed in Mongolia in the 1920s and 1930s. Property, including rights to use pastureland, would thus be confiscated from the princes and lamas and redistributed among the herders, now labeled arad.

"Nomadic feudalism" guided socialist-era Mongolian policymakers' understanding of the historical relationship between land, herd, and herder. As restated by the twentieth-century Mongolian historian Sh. Natsagdorj, the relationship between land rights (or "ownership") and ownership of other kinds of property, especially herd animals, was inextricably intertwined:

> Although the *arat* [common people] owned cattle, dwellings, and other property, they did not have the right to own land, but could only use it. The economic power of the feudal class consisted not only in the ownership of vast numbers of livestock, but also in the fact that they enjoyed a monopoly in the ownership of land rights, which provided the fundamental and unique condition for the viability of pastoral nomadism. . . . The sources of class conflict in Mongolia [in the feudal era] were pasture and livestock in combination.[19]

In order to leap from feudalism to socialism, Mongolia would have to undergo collectivization, that is, private holdings would have to be voluntarily or coercively refashioned into communally held and state-managed assets. This was the methodology mandated by the Soviet advisers who essentially micromanaged Mongolia's economic and social restructuring throughout much of the twentieth century. Obviously, land had never truly been "owned" by any group in presocialist Mongolia, although, as Natsagdorj suggested, the secular princes (in his vocabulary, "feudal nobles") and religious authorities certainly monopolized the best pastures. Collectivization in Mongolia thus would mean pooling herds and labor, not the land itself. This is not a subtle distinction. In fact, Mongolia's collectivization was unique, as compared to collectivization in the Soviet Union and China, in exempting the land itself from the process of reallocation, since, to a great extent, communal principles already governed access to pasture.

In presocialist-era Mongolia, authority over pastures had generally been exerted through communally accepted norms of usufruct, rather than through detailed legal documents that in sedentary societies are used to define strict terms of ownership. The socialist-era negdels did not prevent herders from nomadizing with their herds from one seasonal pasture to another. The herders did, however, become salaried state employees subject to layers of state and party (Mongolian People's Revolutionary Party) bureaucratic administration.

One of the greatest changes in the period 1959–1990 for herders was the mandate to specialize in one type of herd animal, rather than raise multispecies herds that had traditionally been common throughout Mongolia. Just as Wall Street investors are often advised to diversify their portfolios, Mongolia's nomadic encampments (*khot ayil* or simply *ayil*) customarily had relied on three, four, or all five of Mongolia's so-called five animals (*tabun khoshighun mal*) in order to reduce risk of disease that might strike one species but not the others. The Mongolian historian N. D. Bumaa recounted her mother-in-law's specialized work in a negdel: "She recalls that she had to wake up very early to milk many cows so that her cooperative could meet its government-imposed quota and she could receive her salary."[20]

The negdel itself consisted of subdivisions of smaller work units called *brigad* ("brigades" from the Russian word) and *suuri* (even smaller work units that often consisted of eight or fewer households). The suuri would herd only one type of livestock, that is, only horses or only camels. The Mongolian People's Revolutionary Party leadership drew up this blueprint of specialized control as well as the five-year plans that set production goals for herders to achieve. This external source of authority reduced herders to state employees without much scope for independent decision making vis-à-vis pasture use or other issues. As several observers have pointed out, the resulting unquantifiable loss of know-how among herders contributed to disastrous outcomes in the 1990s with the sudden end of the negdel system.[21] In fact, Namkhainyambuu wrote a manual for herders in the 1990s to address the significant problem of reduced skills in such areas as seasonal pasturing, breeding, and restocking of livestock.[22]

The socialist-era doctrine of specialized livestock herding led to some egregious errors, such as segregating sheep from goats. Sheep and goats have been herded together in Mongolia and throughout other pastoral regions of Inner Asia from earliest times and for good reason. The practice is based on certain reciprocal advantages: sheep tend to overstay in any one pasture, but the more mobile goats keep the sheep on the move. If goats are not surrounded by sheep in the colder months, goats are prone to suffer from the cold and from malnutrition. Sheep hooves are better suited than goat hooves to scraping snow off edible winter pasture.[23] As one field researcher notes,

"Because sheep and goats can learn from one another to better survive the harsh environmental conditions, it is beneficial to structure the herd in a cross-species, cross-generational herding structure."[24] Thus, the socialist-era segregation of sheep from goats was not a successful experiment.

Another attempt at modernizing Mongolia's livestock sector revolved around introducing imported herd animals in the 1960s. Unfortunately, unlike the traditional Mongolian breeds, imported livestock were less resistant to the cold and demanded more hay and protection from the winter elements.[25] Crossbreeding with Russian livestock created "super breeds" that depended upon supplementary fodder supplies.[26] In the 1990s, with the end of Soviet subsidies for negdels, supplementary fodder was often unavailable. Ultimately, nonindigenous breeds could not adapt well to Mongolian climatic and environmental extremes. Indigenous breeds proved to be hardier.

A further example of Soviet socialist ideology dictating how herders should live was the attempt to move at least some herders into concrete buildings. Reportedly this effort at sedentarization resulted in loss of livestock to wolves and rustlers, since herders could not hear their herds and flocks as they normally could through the felt covers of a *ger*. Without being able to hear livestock vocalizations at night, herders could not intervene when livestock were threatened.[27]

In 1965, the number of negdels stood at 289, with each one consisting of some 500 herding households on average.[28] In geographic terms, the negdel unit was virtually identical to the *sum* or county, and thus one encounters the administrative term *sum-negdel*. Each negdel (or sum) was organized around an administrative center where administrative, economic, cultural, and health services were located. Just as decision making over such issues as pasture rotation and herd composition was centralized in the hands of local negdel and sum officials, so economic planning overall was centralized, with quotas set by higher bureaucratic echelons and transmitted to local leaders to broadcast to herders who had limited input into the process. Herders were organized into the herding units called brigad, and, as outlined above, the brigad were again subdivided into smaller units called suuri. These smaller units replaced the traditional khot ayil, but were similar in size with two to five households. In one sum in Khovd Aimag, for instance, researchers found that suuri existed only on paper and that the brigad functioned as the smallest unit of herding households.

The negdel and sum leadership was responsible for implementing the state's five-year plans and fulfilling local quotas set by those plans. Namkhainyambuu, in his autobiography, gamely accepted rewards and accolades for meeting and overfulfilling quotas in raising sheep, and demonstrated that a hardworking herder could do rather well in the negdel system. Yet even the loyal Namkhainyambuu could complain about having to participate "in all these Party, State, and public organizational meetings and conferences."[29] The herder's patience could run thin as the demands of negdel bureaucracy consumed time that could have been spent tending to animals. As Namkhainyambuu bluntly stated, "To force the herders to fill in charts and write memoranda and reports is a waste of time."[30] Incidentally, Namkhainyambuu could write a bit more freely expressing such minor discontents because his autobiography was published in the fading twilight of the socialist era, 1989.

Certainly, Mongolian herders enjoyed many benefits of the enormous Soviet subsidies: mechanization of aspects of herding made the pastoral nomadic life easier in many ways. For instance, the negdels could supply trucks and other motorized vehicles to herder families for their seasonal moves from one pasture to another. As Namkhainyambuu noted, "We use draft animal transport less and less, and it is only during the spring and summer migrations that the camels are loaded. Most of the time, cars and tractors are used."[31] Hay mowing equipment and hydraulic wells were all financed by Soviet subsidies to Mongolia in the socialist era. These advances were of course welcome, but Namkhainyambuu realized that superficial modernization could turn people away from their traditional crafts, which might have been superior in quality:

> Industrial-made leather goods, saddles, and bridles don't stand up to the test of the bucking and kicking of wild, unbroken horses... Wooden tubs and pails for carrying one's personal drinking water have been replaced today by various plastic and iron pails which rust out.[32]

Yet, the negdels did provide services that had previously been nonexistent in rural Mongolia. Among the most important state-subsidized services were veterinary care, winter shelters for livestock, mechanical wells, and state-managed delivery of emergency fodder when zud occurred. In the socialist era, each negdel was assigned a veterinarian; by contrast, a decade after the demise of the negdel system there was one veterinarian on average for some 40,000 herd animals.[33]

Herders' cultural horizons were expanded by the introduction of communal recreation and reading centers that were established in the brigade centers of every negdel. These community centers were modeled on the Soviet *krasnyi ugolok* ("red corner"), a term that was derived from the icon corner of a traditional Russian home. The community center often contained, or was adjacent to, a library with books and magazines, many of which had been translated from Russian into Mongolian. While the reading material served as a conduit for state and party propaganda, herder literacy rates nonetheless skyrocketed. In 1960, 545 of these community centers existed in the Mongolian countryside; by 1980, the number had increased to 1,362. Libraries in or attached to these centers grew in number from 33 in 1960 to 772 in 1980.[34]

Cultural outlets were better funded in Mongolia's countryside in the socialist era than they are nowadays. Typically a sum-center library today will have rather out-of-date reading material. The French sociologist Linda Gardelle interviewed a young herder in the Bulgan region who gave this description of the decrepit sum-center library offerings:

> The books are old. You can't get any recent publications. I have a library card but I don't go there often now since there are never any new books.[35]

In addition to the print media, the socialist-era government of Mongolia made sure that radios could serve as a means of propagandizing current policies and priorities. According to V. V. Graivoronskii, who conducted extensive fieldwork among herding households throughout Mongolia in the socialist era, practically every herder household had its own radio. In 1980, four out of every five families in Mongolia (including urban families) regularly listened to the radio.[36] Unlike the radios of the socialist era that were distributed in order to transmit government and party

Figure 4.1 Dismantled community center near Ölziit Sum, Dundgov' Aimag, 1996. Many such buildings in rural Mongolia were stripped for materials by locals once Soviet subsidies for *sum* centers disappeared after 1990.

policies to every herding household, today's evermore common televisions allow herders to be well informed and to formulate their own opinions about political parties, current parliamentary agendas, and other issues.

In terms of Mongolian culture under socialism, the socialist era witnessed not only the persecution of Buddhism (as noted earlier in this chapter) but also the attempt to eviscerate shamanist and animist beliefs. It was far more difficult for the socialist authorities to alter long-standing beliefs that mountains, specially sited lone trees, or other unique landscape features were endowed with spiritual powers. The physical survival of *ovoo*, for instance, provides an interesting example of spirituality in the socialist era.

The use of ovoo as pasture boundary markers in the Mongolian countryside was mentioned in chapter 3. Ovoo also embodied a spirituality connecting humans to the landscape itself. Many scholars have noted this core aspect of traditional Mongolian culture. Most recently, the anthropologist M. A. Pedersen succinctly expressed the role of the ovoo:[37] "Across the Mongolian cultural zone, *ovoos* have since time immemorial been constructed at places believed to be the abode of 'land masters' (*gazryn ezed*), spiritual entities that are considered responsible for the natural conditions (such as rainfall, disease, and fertility) on which human and animal life are dependent."

As Pedersen has noted, in the socialist era, the tens of thousands of ovoo in the Mongolian countryside were not destroyed, but rather were "rebranded" as road markers.[38] Ovoo had indeed always served a mixed purpose as both spiritual sites where travelers seeking safe passageway would circumambulate the ovoo cairn three times clockwise adding a stone each time and as practical markers for pasture boundaries. In the socialist era, however, the authorities attempted to erase the spiritual dimension. The following anecdote may shed some light on this. In September 1978, while visiting Khara Khorum, the earliest capital of the thirteenth-century Mongolian empire, I asked the Juulchin (state tourist company) guide about the sole surviving aboveground relic of the thirteenth century, a stone tortoise that had long ago served as the base for a stele. Specifically, I inquired why there were so many small pebbles and stones atop the tortoise's broad back. Of course the stones had been placed there as offerings by Mongolian visitors in the same manner that the Mongols add stones and pebbles to an ovoo. The guide strode up to the tortoise and swept the stones away with his hands, explaining that some people unfortunately still believed in old superstitions. In 2002 and 2006, however, I found the tortoise happily decorated with blue *khadag* (prayer scarves), which are often left on ovoo and other sacred sites.

While collectivization introduced both positive material benefits and negative bureaucratic elements into herders' lives in Mongolia, pastureland itself continued to be collectively managed as open range without fences. No push toward private ownership of land was ever attempted by the MPRP. Mongolia's traditional open access to pastureland was economically and ideologically acceptable to the country's socialist planners. Instituting fencing, privatization of pastureland, and sedentary-style ranching would have been enormously costly. The Soviet Union's subsidies to Mongolia's civilian economy were largely devoted to urban industries, mining, and transportation (particularly railways to export Mongolia's mineral wealth to the USSR). Thus, for financial reasons, pastoral nomadism was allowed to continue in many ways along traditional lines, with the exception of herd composition.

The USSR mirrored the Qing Dynasty in this regard: each imperial power avoided disturbing a deeply ingrained way of life in Mongolia in order to achieve other, more important, goals including military ones. The Sino-Soviet split that emerged publicly in the 1960s, though it had long been brewing, led to the USSR stationing massive numbers of Soviet troops inside Mongolia's sovereign borders; the Soviets also built missile bases in Mongolia. By 1979, some 120,000 Soviet soldiers were stationed in Mongolia in order to face off against China should the need arise.[39]

Ideologically, it would have been rather contradictory for MPR state planners to have privatized pastureland, a traditionally communal form of resource-sharing. After all, socialist ideology in practice had found its target: the redistribution of herd animals. The land was already communally held. Thus, while herd structure was indeed subject to radical restructuring and herders themselves were taught to regard themselves as state employees on salary, the land itself escaped the socialist-bureaucratic impulse toward meddling.

The redistribution of herd animals by the negdels fulfilled the socialist agenda of leveling wealth. Among pastoral nomadic herders, wealth was defined by the number of herd animals, not by land, dwellings, or other assets as in sedentary societies. According to Goldstein and Beall's fieldwork in Moost Sum in Khovd Aimag, the negdel authorized pasture usage based on the herd size and herd composition of each household in the negdel.[40] In their designated pastures, the herders were permitted to choose campsites to which they typically returned on a seasonal basis. The negdel authorities, however, decided the schedule of nomadic migrations from one seasonal pasture to the next.

The negdel leaders and those at higher levels of state and party bureaucracy were resented by herders at times for micromanaging and even hindering the herding way of life. Herders could judge themselves as far more expert in the techniques of livestock herding and pasture rotation than visiting bureaucrats. Writing in the last years of the socialist era, the herder-writer Namkhainyambuu pithily expressed this sentiment:

> From nation to *aimag*, from *aimag* to *sum*, and from *sum* to brigade, the representative travels by car on his/her cherished mission, wasting

> fuel and oil, getting worn out, and incurring debts... Some herders are scolded and handled unpleasantly by those acting officially who do little to help us.[41]

Namkhainyambuu, who valued learning through experience, castigated the specialists of the socialist era who attempted to superimpose methodologies and production schedules upon herders:[42] "Today there are a lot of representatives who don't know about herding but they teach herdsmen, and they grandly announce plans of work to be done within a definite period of time."

Namkhainyambuu wrote his autobiography in the late 1980s when the Mongolian version of *glasnost'* emerged. Thus, when addressing the somewhat sensitive issue of private herd animals, a small number of which were permitted to each household, he could voice the opinion that "the question of private herds should be rethought, and the restrictions on personal large livestock herds must be stopped."[43] It was not that Namkhainyambuu was anti-collectivization; rather, he seems not to have perceived a contradiction between the herder as wage-laborer taking care of the negdel's livestock and the herder as private owner. Private herd animals, whose numbers were strictly limited by the state until the late 1980s, generally supplied herding households with basic dairy and other needs, and thus the private herds maintained the traditional, presocialist multispecies composition. In Moost Sum in Khovd Aimag, for instance, a herder who might be assigned to tend only female yaks from the negdel's herds, would likely maintain a five-species (horse, camel, yak, sheep, goat) private herd.[44] In the late 1980s, the limits on private herd numbers were raised and finally in 1990 all such limits were eliminated. The socialist era was coming to an end, and it would soon be up to the herders to determine which aspects of the negdel system to maintain, which aspects to change or abandon, and finally what to implement in the newly created absence of a strong state structure.

As one tallies up the advantages and disadvantages of the socialist-era negdel system—from a herder's point of view—one could view positively the artificially propped up economy of the MPR: Soviet subsidies allowed for little incentive for negdel leaders to punish herders who failed to fulfill quotas, which were generally set rather low anyway. From a broader point of view, however, the resulting stagnancy in productivity was readily apparent. In 1940, livestock numbers rose to about 26 million, and from 1945 to 1990, livestock numbers fluctuated between 22 and 25 million head.[45] Incentive in the socialist era was directed at rewarding self-motivated herders like Namkhainyambuu, who in his autobiography, proudly described being the recipient of several state awards and medals, such as Aimag Champion, Hero of Labor Award, the Star of Sukhbaatar, and the Golden Soyombo. Herders less inclined to put out effort, however, could coast along, confident that the negdel would pay roughly the same wages regardless of their output. The herders' households would never suffer hardships with the socialist safety net firmly in place.

With the abrupt end of Soviet subsidies to Mongolia in 1990, the herders would face a jolting reality. Even the best personal work ethic would not necessarily be sufficient to feed one's household much less accrue profit. Economic forces beyond the ken of the herders would wreak havoc with the traditional pastoral nomadic economy for the better part of a decade and beyond. Just as the Mongolian state would be given a potentially bankrupting bill for national debt in the amount of 11 billion dollars—payable to Russia—the herders themselves would suddenly face the reappearance of actual risk with its severe consequences in the pastoral nomadic way of life. Finally, in late 2010, two decades after the Soviet Union had ceased to dictate policy to the MPR, Russia wrote off 97.8 percent of Mongolia's remaining debt, which had already been negotiated down, leaving only 3.8 million dollars for Mongolia to repay. This put to rest the largest fiscal issue left over from the socialist era, but the Mongolian herders' problems could not be magically waived away by governmental fiat.[46]

While the negdels would be eliminated structurally and physically after 1990, as we shall see, there have been significant carry-overs from the socialist era that continue even today to influence work habits, expectations of government, and other social attitudes among Mongolia's population at large including herders.

A small but evocative example of the lingering culture of socialism surfaced in the aftermath of the devastating zud of winter/spring 2010, when the government of Mongolia's National Emergency Agency in conjunction with the United Nations Development Programme (UNDP) presented awards to 54 herders who demonstrated preparedness and thus lost no more than 10 percent of their herds to the effects of the zud. The awards ceremony at the Government House in Ulaanbaatar which 12 of the herders attended in June 2010 was eerily reminiscent of the socialist-era ceremonies, which Namkhainyambuu, the recipient of many such awards, so contentedly described in his autobiography.[47] The Mongolian commentator, G. Chingis, in 2010, labeled the ongoing custom of excessive awarding of titles and medals part of a "huge Russian heritage" from the period of Soviet domination over Mongolia. In a somewhat facetious editorial, G. Chingis added that "the Russian invasion of our mentalities has been much more effective and victorious than our actual numerous invasions of their homeland."[48]

Another Soviet-era legacy that continues to influence modern-day Mongolia relates more to urban than rural life, but still reverberates throughout society. After 60 years of socialism that saw the Soviet Union's push toward creating an urban work force in Mongolia, the country still has an insufficiently trained and motivated work force. Robert Rupen, writing in 1979, analyzed some of the reasons behind the low productivity, high absenteeism, and other work-related problems among Mongolia's urban workers under socialism:

> Adding to the problems caused by the labor shortage are widespread apathy and significant motivational and attitudinal problems. The regime blames these on feudalism and Buddhism, but they may actually be the results of the traditional nomadic life-style of the Mongols, of Communist economic forms and organization, and of the laborers' lack of interest in working hard to satisfy Mongolia's debts to the USSR.[49]

Mongolian workers were mindful of the fact that the copper and molybdenum mined at Erdenet in Northern Mongolia were transported via rail to the USSR. Rupen and other scholars have used the term "neocolonialism" to describe Mongolia's economic status: the USSR largely subsidized the MPR economy and provided expertise and trainers to guide and motivate the less than enthusiastic Mongolian workforce. Why bother to work hard to enrich the elder brother to the north?

G. Chingis, writing for *The UB Post*, has taken up the issue of Mongolia's socialist heritage and its ongoing hold over Mongolian politics and economics. He writes that "it is extremely difficult for Mongolians to leave our socialist heritage, especially when our country is surrounded by two socialist countries—The Russian Federation and The People's Republic of China."[50] Ironically, notes the editorialist, Mongolia's neighbors are attempting to escape the socialist model (this is somewhat debatable in the Russian case), but Mongolia's political leaders have much to gain by keeping or creating state-owned companies in which many of them own shares.

Yet, among herders and others, a nostalgia for the Soviet-dominated era of modern Mongolian history quickly developed as the economic transition from a socialist centralized economy to a free market economy led to grim outcomes for both rural and urban dwellers in the 1990s. As the anthropologist David Sneath noted in 2003, "The real transition that Mongolia has experienced has been from a middle-income to a poor country, as if the process of development had been thrown into reverse."[51] With the dismantling of the socialist infrastructure and withdrawal of subsidies from the Soviet Union, all sectors of Mongolian society after 1990 would face undreamed of challenges. Five-year plans and other staples of socialist economic planning were swept away as new concepts of the free market and capitalist enterprise were abruptly introduced, often with severely negative short-term impacts for herders as well as urban dwellers. The issue of land use—to privatize pastureland or to continue customary shared usage—was only one of many that would take decades to work out.

The following chapter will examine the dismantling of the socialist negdel system, the reallocation of livestock and other collectively held assets such as winter livestock enclosures, and the evolving debates over pastureland usage.

CHAPTER 5

POST-1990 DEVELOPMENTS: WHO DETERMINES LAND USE IN THE ERA OF THE FREE MARKET?

Land use in post-1990 Mongolia has been in constant flux, often reacting to environmental, demographic, economic, political, and legal pressures. This chapter will attempt to trace the changes in land use and examine ways in which herders participate in decisions regarding land use issues in Mongolia's countryside. Land use in post-1990 Mongolia cannot be summarized easily: variation in practice belies the notion that laws enacted by Mongolia's *Ikh Khural* (parliament) will lead to uniformity and fairness across the landscape. This chapter will also place Mongolia in a comparative context, examining policies governing herders in the pastoral zones of the People's Republic of China and Kazakhstan. The role of agriculture in Mongolian history and agriculture's future in Mongolia are important issues that will be addressed in chapter 6.

PEACEFUL TRANSITIONS FOLLOWED BY ECONOMIC CHAOS

In the winter months of 1989–1990, a peaceful transition from a socialist one-party political system to a democratic multiparty parliamentary system took place in Mongolia's capital city, Ulaanbaatar. In stark contrast to the 1989 Tiananmen disaster in China, where pro-democracy demonstrators, supporters, and bystanders were massacred by the People's Liberation Army on orders of the Chinese Communist leadership, the Mongolian People's Revolutionary Party (MPRP) stepped back from power, acceding to the demands of hunger strikers on Sukhbaatar Square.

With the evolution of a new parliamentary system and the birth of new political parties came a new state constitution (1992) and, in the years following, an array of laws and regulations directed at land use, land registration, and land ownership. Examining the events surrounding the dismantling of the *negdel* system, it is hard not to conclude that the good as well as the bad were all hastily thrown out before new social and economic structures could be constructed. This was particularly obvious in the countryside, where negdel assets were unevenly and often unfairly redistributed. The scramble for booty was a common feature of the fall of communist regimes in the Soviet Union and in all of its client states.

Ideally, the privatization of the negdel meant the redistribution of all negdel assets including herd animals. The 1991 Law on Privatization directed all Mongolian citizens born before May 31, 1991, to receive vouchers that would enable them to purchase previously state-owned property. Pink vouchers were intended for smaller properties (in the countryside, this would include herd animals and winter-spring shelters) and blue vouchers were intended for larger enterprises such as factories or negdels.[1] Because of the suddenness of the transition from top-down centralized planning to individual or household-based decision making, many herders chose to transform their disbanded negdels into shareholding companies in 1990–1992. With the unknowns of the free market, it made sense for a herder to avoid the enormous risks of going out on one's own after decades of guaranteed state purchase of the products of the herding economy.

In Melvyn Goldstein and Cynthia Beall's case study of Moost in Khovd Aimag in western Mongolia, all but 2 percent of the former negdel members chose to join the shareholding company. Goldstein and Beall quote a local official's insight into why so many herders chose to stick together in the early 1990s:

> The herders decided to remain in the company for the security. They are waiting to see how the company develops and also how the private

> herders manage. They were not clear how they would market their products independently and many were not convinced that Mongolia would really become a complete market economy, so they hedged their bets.[2]

Initially this seemed to be the least risky option given that the new companies were structured along the lines of the former negdel and the leadership remained the same. Goldstein and Beall optimistically concluded that "this first phase of the transformation of herding collectives has worked well."[3] As it turned out, however, the shareholding company was a short-lived step in the complex decollectivization process; most such companies had gone bankrupt within a few years.[4] The absence of state subsidies and state procurement and the nomads' unfamiliarity with marketing strategies all contributed to the failure of shareholding cooperatives. Even though the traditional *khot ayil*, a long-standing Mongolian social unit defined by Robin Mearns as "a fluid group of herding households that co-operates in livestock and pasture management, notably to take advantage of labour economies of scale,"[5] was revived, it could not fill the vacuum left by the former negdel system in managing pasture access.[6]

With the disappearance of a state safety net, many herding households fell into poverty, and a wide disparity between rich and poor herding households quickly emerged. According to a 1992 estimate, 59 percent of rural households owned fewer than 50 animals apiece, and 19 percent had fewer than 10 animals apiece.[7] These figures certainly suggest barely viable livelihoods. Pressure on pasturelands increased as a direct result of the relocation of unemployed urban dwellers to the countryside: the number of herding households doubled between 1990 and the mid-1990s.[8] There is also evidence of spatial rearrangements among herding households themselves owing to impoverishment. Martha Avery's volume, *Women of Mongolia*, offers glimpses into women's lives in Mongolia in the early 1990s. One of her interviewees, a destitute young woman living in the Gobi with her child, provides a striking tableau of relocation through necessity:

> I've been living here for four years. We came from further south in the Gobi. It took five days. We came by camel caravan, my brother-in-law, my sister, my family. The land was not as good down there, and we knew about this place. Nobody had settled here, so we moved in.[9]

Avery's comment on this woman's situation speaks more broadly of the climate of the early 1990s:

> In most parts of the country there is no apparent restriction on occupying unused land. Squatters' rights appear to prevail. International consultants are busily debating land-use laws for Mongolia. Meanwhile, the structures and restrictions of the previous Soviet regime have gone and people are moving, positioning themselves for the future.[10]

Land Legislation, the Impact of Zud, and a Few Comparisons across Borders

As David Sneath has pointed out, pressure in favor of land registration and land titling was already building in the early 1990s, particularly among economists at the Asian Development Bank.[11] While the resulting 1994 Land Law, which went into effect in 1995, did not go so far as to authorize private ownership of land, it did allow leasing of winter campsites and pastures to individuals.[12] The 1994 Land Law allowed great discretion to local officials in resolving the particularly important issue of rights over winter campsites and winter pastures that are vital to the survival of the nomads' herds. The 1994 Land Law was ambiguous on the hierarchy of authority involved in adjudicating conflicting claims over pasture and campsites, although the onus seemed to rest on local officials.[13] Anecdotal evidence from the early postsocialist period suggests that herders perceived government officials as disinterested or too passive on issues of importance to herders. Even a relatively well-to-do herder with more than one thousand head of livestock felt abandoned by the government:

> It seems to me that the State does not pay much attention to the hardships and labours of herders. It does little except the selection of the year's best herder at the end of the year.[14]

How could the promise of maintaining common and free access to pastureland in Mongolia—a promise explicit in both the 1992 Constitution and the 1994 Land Law—be reconciled with the decidedly different approach to land as a marketable commodity in the 2002 Land Law?[15] First, one must understand that the 2002 Land Law, which went into effect in May 2003, was aimed primarily at urban land ownership. The 2002 Land Law was followed by other urban-focused legislation such as the Law on Land Ownership (effective May 2003) and laws on housing and apartment privatization.[16]

Yet, during the parliamentary debates leading up to the passage of the 2002 Land Law, great concern was voiced over the implications of this pending legislation for land use and claims over land in rural areas. One area of concern lay in the potential misuse of authority by *sum* (county) and *bag* (district) officials in allocating land rights to herders; wealthier herders would, it was feared, be given preference thereby exacerbating inequalities among herding households.[17] The provision of thc 2002 draft land law under discussion in June 2002 calling for fee payments for land use including use of pastures by herders proved very controversial.[18] On June 28, 2002, the then prime minister Nambaryn Enkhbayar defended the new land law on television and reportedly assured the Mongolian people that public-use lands, including pastureland, would not be covered by the law. At the time that the 2002 Land Law was passed by the Mongolian parliament, only 0.9 percent of Mongolian territory was to be subject to private ownership according to the new law, while agricultural and pastoral nomadic land use issues were put off for future debate and regulation.[19]

The 2002 Land Law led to 90 percent of apartments in Ulaanbaatar being privately owned (although the urban land under the apartment buildings was still state owned), and also led to large-scale privatization of *ger* settlements within the city's precincts.[20] Given that the impact of the land legislation had been primarily urban in nature, why then would Mongolia's herders in the early 2000s worry about the law's extension into pastoral nomadic lands? One reason stemmed from the disastrous *zuds* of 1999–2001, which began with drought in summer 1999 and continued through two winters of heavy snow (1999–2000; 2000–2001) and resulted in the deaths of an estimated 6–7 million head of livestock in Mongolia. Several commentators argued that disasters on such an immense scale would not have occurred in the socialist era of central state planning when the negdels had the capacity to distribute stockpiled hay to needy herding households in emergencies.[21] In the aftermath of these back-to-back zuds, the viability of herding per se was called into question, with some local officials expressing more confidence in mining as a source of revenue.[22]

In 2001, Mongolia's then prime minister Nambaryn Enkhbayar called for the end of nomadism, partly in reaction to the two seasons of zud, and partly as a means to expedite his radical, utopian plan to urbanize 90 percent of Mongolia's population within 30 years. In a May 2001 interview, Enkhbayar declared: "It is not my desire to destroy the original Mongolian identity but in order to survive we have to stop being nomads."[23] Again, in 2002, Prime Minister Enkhbayar in an interview with BBC predicted that the nomadic way of life in Mongolia would disappear within 10–15 years. He again portrayed the pastoral nomadic segment of Mongolia's population as backward and as posing a hindrance to national survival.[24] Nomadism has periodically been lauded by Mongolia's leaders as part of Mongolia's national identity while paradoxically at other times it has been labeled "backward," depending upon the audience at hand.

Is the "Tragedy of the Commons" Applicable to the Grasslands of Mongolia and China?

The disastrous zuds of 1999–2001 also played into the hand of those arguing that private land ownership would improve the lot of the herders who, in the wake of catastrophic losses, were in dire need of economic assistance.[25] Many Western economic aid agencies continue to abide by the theory of the "tragedy of the commons," an influential theory that argues that overuse and degradation of common-access land is inevitable in the absence of private landownership.[26] Important research over the past 20 years, however, has convincingly challenged this theory: common property arrangements do not necessarily lead to environmental catastrophes, particularly when empowered groups of users adopt and enforce rules in the common interest.[27] While many economists favor privatization of communal resources in order to solve problems of overuse, Elinor Ostrom and other scholars of common pool resource management have developed strong arguments based on field research showing that self-governing institutions can effectively regulate the use of communal resources—like pasture and water. Nonetheless, the so-called tragedy of the commons idea meshed well with Western aid organizations that from the

1990s onward have generally favored privatization as the panacea for Mongolia's economic woes. For example, in 2003, a United Nations Development Programme (UNDP) report stated as fact:

> In Mongolia, the unregulated availability of grass—the primary input in the production of animal husbandry—is already leading to the classic "tragedy of the commons", with the degradation of the shared space.[28]

A similar line of argument has been pursued by the government of China in the aftermath of climatic disasters affecting Tibetan nomads. The Tibetan example is instructive. Daniel J. Miller examined the Chinese government's response to the loss of some three million head of livestock in the Tibetan Autonomous Region (TAR) owing to heavy snowfalls during the winter of 1997–1998. In spite of the aid provided by the Chinese government and relief agencies, it would take many years to rebuild the devastated herds. Yet, Miller points to more long-lasting reverberations in the fields of economic-environmental policy and social engineering:

> The livestock losses on the Tibetan plateau following the heavy snows are now seen by many officials as proof of the need to settle Tibetan nomads and to introduce more "modern" and "scientific" animal husbandry practices. Government officials in China generally believe that nomads are backward and that their traditional livestock and grazing management systems are an improper use of the land. Nomadic pastoralism is thought to lead to over-grazing and rangeland degradation... Most officials in China insist that, for development to be achieved in Tibetan nomadic areas, nomads must be settled, houses and barns should replace the traditional nomad yak hair tent, rangeland must be divided into individual family units and fenced, herds need to be restructured, livestock numbers should be adjusted to carrying capacity, fodder has to be grown for the winter, and, for the rangelands, "ecological engineering" and "grassland construction" needs to be undertaken.[29]

Miller's research challenges Chinese governmental assumptions about the viability of traditional Tibetan pastoral nomadic practices. As Miller points out, severe cold and snowstorms are recurrent, natural phenomena in Tibet's ecosystem, and livestock losses similar to those of the harsh winter of 1997–1998 have occurred periodically over thousands of years on the Tibetan plateau, with herders nonetheless surviving and eventually replenishing their herds. David Sneath has made the same observation regarding the recurrent zuds in Mongolia, referring to such natural disasters as "a feature of life on the Mongolian steppe."[30] This is not to suggest that Mongolian herders in the era of the market (*zah zeeliin üye*)[31] should be left adrift with no assistance, nor is it to suggest that the enormous cost of restocking should land by default in the Mongolian government's hands. What other options exist?

An ambitious plan to insure the livestock herds of Mongolian herding families shows great promise for future nationwide implementation. This "Index-Based Livestock Insurance Project," (IBLIP) implemented in a few pilot aimags through the Mongolian Ministry of Finance and Economy and the World Bank, promises to provide herders with strategies and incentives to manage risk and protect their herds from periodic climatic disasters.[32] Originally confined to a pilot program in four *aimags*, the IBLIP received further funding from the World Bank in early 2010 that will allow the project to expand to all 21 of Mongolia's aimags by 2012. Borrowed from an agricultural model and newly applied to livestock herding, this insurance strategy works by having participating herders contribute to a collective fund that is managed by Mongolian insurance companies. Based on the aggregated sum livestock mortality data collected by Mongolia's National Statistical Office (NSO), insured herders receive indemnity payments without having to prove their individual losses. This innovative attempt to involve herders more actively in determining their own financial futures may also dissuade those naysayers who see pastoral nomadism as a cultural relic with no potential for economic growth. The government of Mongolia in postsocialist times has encouraged World Bank funding in this area; most importantly, the Mongolian government has not started down the path of social engineering that we see in the case of the Chinese government's handling of Tibetan herders.

The theory of the tragedy of the commons certainly has its advocates, yet specialists on pastoral land management have long argued against the theory's applicability in the grasslands of Central Eurasia.

In the case of Tibet, for instance, Daniel Miller has persuasively made the case against the theory:

> This theoretical "tragedy of the commons" problem contends that when many individuals graze their livestock on communal land, it is in the interest of every herder to keep increasing his livestock numbers. It was asserted that private ownership, by combining interest in both land and livestock, would prevent overgrazing. This model has been widely rejected by most pastoral specialists throughout the world, who have found it a very poor guide to understanding traditional nomadic pastoralism and for planning development in pastoral areas.[33]

The grasslands of Tibet have long had the twin advantages of remoteness from administrative centers and vastness that preclude an enormous investment in fencing. Yet, in more recent years, the Chinese government has increased efforts to settle Tibetan nomads into villages in the TAR as well as in Qinghai, Yunnan, and other ethnically Tibetan regions. Some of these efforts to remove Tibetan herders and farmers have reportedly been undertaken in order to free lands for redevelopment or to "improve the livelihoods of pastoralists."[34] In addition to the Chinese governmental and popular perception that urban life is inherently superior to rural life, the notion of restoring pastureland by removing livestock herds has fueled various campaigns on the Tibetan plateau to move herding households into urban settlements. Tens of thousands of herder households in ethnically Tibetan regions of Qinghai are slated to be moved into resettlement villages that have been dubbed "theft schools" owing to the poverty and high unemployment in those government-subsidized outposts that already exist.[35]

In terms of climate change and its effects on the grasslands of the Tibetan plateau, recent research strongly suggests that grazing of livestock may actually buffer pastureland from the effects of global warming trends. Grazing of livestock for instance may prevent losses in plant diversity that would otherwise occur with the warming of the ecosystem.[36] Certainly, Chinese government efforts to solidify property boundaries and decrease the range of herder mobility have eliminated for Tibetan herders the crucial option in severe weather—like snowstorms—to migrate temporarily to better pastures with less snow cover. In other words, Tibetan herders in China can no longer go on *otor*, the Mongolian term for a temporary, long-distance migration to avoid zud. The boundary lines demarcating pasturelands in Qinghai and in Inner Mongolia have tightened up, leaving herders in these regions with fewer and fewer options.

In Inner Mongolia too, one may easily identify the effects of Chinese central government social and environmental engineering. The grave socioeconomic and environmental drawbacks of China's grasslands policies have been documented through Dee Mack Williams's extensive fieldwork and commented upon by David Sneath.[37] With the Chinese state's declared policy of settling the nomads of Inner Mongolia and with the introduction of the practice of leasing pastures starting in the 1990s, nomadic herders in Inner Mongolia have had little or no incentive or ability to rotate pastures or to undertake longer, seasonal migrations in search of fresh pasture. The elimination of herder mobility has directly led to pasture degradation. Conflict over fencing seems to have supplanted conflict over pastures per se in Inner Mongolia, and pasture enclosure by fencing has widened the gap and exacerbated relations between wealthy and poor herders.

Even the few remaining reindeer-herders in a remote region of Inner Mongolia have not escaped notice by the Chinese government and have been subject to the forced resettlement policy.[38] With the pretext of conserving the environment and aiding the social development of an ethnic minority, the Chinese government in 2003 relocated Evenki reindeer-herders from the taiga zone of northeastern Inner Mongolia into a newly constructed settlement. As is typical of state-nomad relations in modern Chinese history, the state accused the reindeer-herders of "environment degradation" through herding and hunting, while ignoring state-subsidized timbering and industrial development (which are presented as rationally planned, modernizing development) in the same region. The resettled Evenki reindeer-herders expressed their sentiments about their ruptured way of life to a Western anthropologist with statements such as:

> Of course life is harder in the forest, but it is the place where our culture survives. This is why we withstand the hardships...It is nice to have a house and I am happy my son can stay there after school, but the settlement is like a zoo.[39]

Evenki song lyrics also bespeak great loss:

> I sing about my love for the forest and my reindeer,
> How can I sing such songs when I cannot see the stars?
> We may not see cars and trains,
> But we are happy because the forest is our home,
> For us, life without reindeer is like life without rice for the Han.[40]

Thus, the examples of Tibet (including ethnically Tibetan regions in the provinces of Qinghai and Yunnan) and Inner Mongolia are highly suggestive of the role that state authority can play in determining herders' access to pastureland. Settlement policies vis-à-vis nomads are not new developments historically speaking. In broader terms, most sedentary imperial administrations throughout Eurasian history have sought to settle nomads and transform them into agriculturalists. This is true not only of the Manchu Qing dynasty (in Inner Mongolia), but also of the Tsarist Russian and Ottoman Empires.

By comparison, in post-1990 Mongolia, the government may be faulted for too little intervention, but it clearly has not erred on the side of social engineering. Let us return to the Mongolian government's policies and implementation in the realm of land law in the postsocialist period.

PHASING IN THE 2002 LAND LAW

On May 1, 2003, the Land Law that the Mongolian parliament passed on June 28, 2002, went into effect. The privatization of 0.9 percent of Mongolia's territory within two years was the stated goal of the new law. Privatization of lands occupied by households in ger districts of Ulaanbaatar and other aimag centers was part of the initial stage, but about 60 percent of the territory to be privatized was categorized as farmland. Prime Minister N. Enkhbayar pronounced May 1, 2003, a "historical day for Mongolia... Parliament and government passed the law on landownership by Mongolian citizens... in order to build conditions to make citizens become owners of land of their native country and strengthen independence."[41]

It was reported that President Bagabandi, in a meeting with the head of the Office of Land Relations, Geodesy and Cartography, asked several questions pertaining to the actual implementation of the new law, among them, "whether the office has defined the infrastructure of the land, how cadastre mapping is proceeding for land to be privatized, whether the office has drawn up rules and run training sessions, and the process of establishing a national land database."[42] The 2002 Land Law contained specific provisions referring to authority over land classification, particularly Article 8, "Maps of Borders, Names of Geographic Units and Land Classification."[43] The 2002 Land Law also specified the duties of local officials from aimag governors, sum governors, to bag governors in regards to submitting annual drafts of land management plans; the sum and bag governors each were accorded the responsibility of deciding "to grant land for use or possession in accordance with the annual land management plan."[44]

Exactly which provisions of the 2002 Land Law applied to pasturelands used by Mongolia's herders? Article 54, entitled "Pastureland, Its Rational Use and Protection," was directly relevant. The distinction between summer and autumn pastures versus winter and spring pastures was delineated in Article 54.2. The former were to be "used collectively" while the latter were to be protected from out-of-season grazing.[45] Fencing of pastureland was referenced in Article 54.5: "Pastureland fenced for purposes of developing intensive settled livestock breeding or farming of tamed animals can be given for use to citizens, companies and organizations regardless of season."[46] The Ministry of Food and Agriculture's website in 2005 featured a policy statement entitled "Food and Agriculture Policy of Government of Mongolia" that strongly supported "intensified livestock production," and further promoted more herders living in "settled and semi-settled modes."[47] In spite of its stated preference of settlement, this 2005 policy did not lead to population resettlement; the governments of Mongolia and China are quite different in this regard.

Fencing of pastureland again pops up in Article 48.1 of the 2002 Land Law: "If land in possession or in use is not specifically protected by erected fences or posted warning signs prohibiting entering and crossing, any person may enter or cross this land without causing damage to the land."[48] As I noted during my 2010 and 2011 visits, fencing was obvious in peri-urban areas, particularly the expanding exurbs of Ulaanbaatar where vacation homes, tourist hotels, semi-intensive dairy farms, and pastureland intermingle. Away from urban centers there were more permanent structures on the landscape (mostly storage sheds), but fencing was limited to enclosures around tourist ger camps. Based on personal observations in the summers of 2010 and 2011, the absence of fencing for pastureland was the norm in the aimags of Töv, Dundgov', Ömnögov', Selenge, and Uvs. There is no signage pertaining to property rights in the countryside. Thus, the phrasing of Article 48.1 would seem to allude to an imagined time in the future when fencing and warning signs might be more common.

As Robin Mearns has pointed out, one article in the 2002 Land Law managed to simultaneously incorporate references to land possession and communal use.[49] Article 6.2 stated: "The following [types of] land, regardless of whether they are given into possession or use, shall be used for common purpose under government regulation." Among the lands listed are "pasturelands, water points in pasturelands, wells and salt licks."[50]

Herders were given some autonomy in regard to conflict resolution in the Land Law of 2002. Article 54.10 stated: "Disputes arising in relation with use of pastureland shall be resolved by discussing them on [*sic*] Bag Public Khurals [i.e., at district-level public assemblies] based on traditional land use practices and customs of herders. If an agreement can not be reached, the issue shall be resolved by governors of soums [sums]."[51] As is so often the case, a precise definition of "traditional land use practices and customs of herders" is absent. The intent of this provision, however, seemed to be to encourage herders to resolve their own land conflicts; only after this first line of action had failed would county-level officials be expected to intervene.

The 2002 Land Law dispensed some vaguely worded authority to local officials to implement regulations governing seasonal pastureland use:

> Summer and autumn settlements and rangelands shall be allocated to baghs and hot [*khot*] ails [*ayils*] and shall be used collectively. Winter and spring pastures shall be prevented from livestock grazing during summer and autumn, and shall be carefully protected with public efforts.[52]

As we shall see later in this chapter, establishing usufruct rights at winter and spring campsites has been formalized in some parts of Mongolia through the use of "certificates of possession." These certificates, however, have not been uniformly adopted, and the terms of pasture tenure vary greatly from one region to another. With the Draft Law on Pastureland awaiting further discussion and possible action at the fall 2011 parliamentary session, there was some hope that local officials would be given the resources to enforce the authority that they hypothetically have governing land use issues. With parliamentary elections looming in 2012, the feeling among many Mongols with whom I spoke was that the Draft Law on Pastureland would very likely be passed.

WHO DECIDES PASTURE USAGE? WHO ADJUDICATES DISPUTES?

The vague phrasing of certain articles of the 2002 Land Law brings to mind a question that unfailingly pops up whenever a first-time visitor to the Mongolian countryside encounters the near total absence of fencing across the grasslands and deserts of the countryside: "Who decides whose herds get to graze this area?" This same question, phrased a bit differently, was posed to 773 herders across Mongolia in a public opinion survey carried out by the Open Society Forum and published in 2004. The survey geographically encompassed 34 sums in 12 aimags plus the Ulaanbaatar region. The research team asked: "What secures your possession of pastures?"[53] Herder response broke down as follows: 43.4 percent said "a long tradition of possession"; 28.3 percent said "our pens and fences are there"; 15 percent said "there are no guarantees"; 11 percent said "[they are] distributed by the *Soum*"; and 1 percent were listed as giving "other" as a response. The "pens" mentioned undoubtedly referred to winter enclosures for the smaller livestock (sheep and goats), and the fences likely referred to the increasing use of fences in peri-urban areas. Based on these responses, the minimal participation of state and local institutions in determining pasture usage is quite striking.

The idea that "tradition" governs usage may seem unacceptably vague to modern social scientists, but clearly "tradition" represents a complex mixture of customs and practices that constitute a kind of unwritten, de facto way of life. While "tradition" as a catchall phrase may indeed be maddeningly resistant to precise definition, depending upon the perspective of a given herder, the concept should not be dismissed out of hand. One Western researcher, Ole Bruun, has taken just such a dismissive approach to this question, however, essentially reducing herders' social customs as they pertain to pasture usage to a fictionalized creation of Westerners:[54] "Even the concept of 'customary' institutions may be deceptive, as it ignores the fact that the rulers of each historical period to a large extent set the terms for pasture allocation by means of their own structures of power: searching for native institutions is a product of tribalization of the Mongols to satisfy foreign romanticism."

Bruun's fieldwork in Arkhangai Aimag was completed before the 2002 Land Law was enacted, and thus he saw "pasture management . . . in a state of disarray in . . . all rural localities" with "no traditional institutions" emerging to replace the vanished socialist-era negdel.[55] Sum governors were ill-equipped in terms of personnel and vehicles to exert authority in pasture allocation issues in the areas studied by Bruun. Instead, Bruun witnessed approximately 100 instances of wealthier herders building small cabins on summer pastures as a way "to mark territories with good grazing."[56] Hundreds more were collecting the materials to do exactly the same. Natasha Fijn also noted the trend in Bulgan Aimag for herders to build a cabin in summer out of black pine, but she assumed that herders chose to do so because a log cabin in summer is cooler than a ger. Fijn did not relate cabin building to a land grab, as Bruun did.[57]

For this author who has traveled extensively in the Mongolian countryside in the years from 1978 to 2011, the number of structures on the landscape has visibly increased, particularly in the central aimags like Arkhangai and Töv. Whether this amounts to what Bruun terms "an informal privatization in full swing"[58] remains to be seen. If sum authorities are ever empowered to deal effectively with pasture management, they will have to either acknowledge the legitimacy of such land claims on their cadastral maps or else confront the builders of such cabins on summer pastures that are, according to the Mongolian constitution, guaranteed as open access. The "certificates of possession" (*ezemshigchiin gerchilgee*), which will be discussed later in this chapter, govern only winter/spring pastures and their associated livestock enclosures, not summer/fall pastures that are supposed to be open access.

As state authority over pastureland issues during the past two decades has been diminished, understanding Mongolian social organization becomes crucial in the attempt to see how herder society functions in the vacuum created by a lesser governmental presence. In recent times, the most commonly used terminology that reflects herders' organization above the household level is *malchin büleg* or khot ayil. The former has been defined as "loose unions of related and/or friendly livestock keepers";[59] the latter has many definitions including my own preferred "encampment of gers"; "camp communities"; "the primary unit of social organization"; and, most elaborately, "a fluid group of herding households that co-operates in livestock and pasture management, notably to take advantage of labour economies of scale."[60] Herders thus have the capacity to self-organize but they must also deal with aimag and sum authorities on issues of land use and the registration of winter livestock enclosures.

As some researchers have argued, the socialist-era administrative units of the aimag and sum with their territorial boundaries are ill-suited for seasonal herd movement. Today's territorial boundaries of aimags and sums may prevent herders from "legally" undertaking longer migrations with their herds.[61] The problem of a top-down imposed administrative structure with units that circumscribe herder movement is, as we have noted in earlier chapters, a recurrent problem in Mongolian history. The Qing era "banners" or *khoshuun* also limited nomadic migrations from pasture to pasture. Jörg Janzen in his multiyear project, "Transformation Processes in the Rural Areas of Mongolia," carried out from 1996 to 2002, concluded with proposals for "larger administrative units which would allow more long-range, ecologically more adapted pastoral migrations"[62] as well as reliance upon herder communities to achieve economic development goals.

When do disputes break out over land and water use? Disputes over pasture usage typically arise when herding groups remain for too long in one area or when herds pass through lands that are being grazed by other herders' animals. As Janzen's fieldwork in western Mongolia suggests, some herders may lack access to a particular seasonal pasture, usually a spring or autumn pasture, thus leading to a situation of overstaying and overgrazing in another seasonal pasture.[63] Janzen explains: "Migration is supposed to follow a schedule agreed on between the herdsmen groups and the administration. If it is not adhered to conflict can arise during movement through the pasture areas of other livestock keeping groups."[64]

When disputes do occur among herding households regarding the right to use particular pastures, who in fact adjudicates and resolves such disputes? The Open Society Forum's survey of public opinion, published in 2004, permits us to get a sense of how herders themselves prefer to handle such disputes. According to the survey, 42.8 percent of herders answered the question "who can settle disputes over pastures?" with the answer "herders themselves"; 39.2 percent answered "soum and bag governors"; 10.3 percent answered "none of them can settle the problem, we need other methods"; 5.3 percent answered "people respected in their area"; 1.3 percent offered no answer; and 1.2 percent answered with "other."[65] The more highly educated herders much preferred having sum governors adjudicate pasture disputes.[66] Otherwise, the survey did not detect any appreciable differences based on age or geographic region among the 773 herders who answered this question.

What is most striking in the replies to this survey question is the confidence and assumed self-sufficiency of herders who obviously deemed themselves capable of handling pasture dispute issues. Possibly some herders may also have held low opinions of particular sum authorities and therefore may have found it preferable to exclude state authority from conflict resolution issues.

The 5.3 percent who answered "people respected in their area" rather than sum or bag governors presumably had some particular figures in mind. Who might such respected people be? With the revival of Buddhism in Mongolia after 1990 and the end of almost 70 years of state repression of religion, might the "people respected in their area" be Buddhist lamas? At this juncture, it may be worth our time to examine the role of Buddhism in the Mongolian countryside nowadays and the role of Buddhism in other pastoral nomadic regions. The comparative perspective will give some sense of how pasture disputes are handled in other traditionally nomadic regions of Central Eurasia and whether religion operates differently in Mongolia.

Pastureland Conflict Resolution in Comparative Perspective and the Role of Buddhism

Western language literature on grassland conflict, mediation, and resolution in China's pastoral regions (primarily Xinjiang, Qinghai) is limited because of the difficulties of access to conduct fieldwork. A 2001 article on property rights in northern Xinjiang found that in Altay Prefecture, an ethnically Kazakh pastoral region where the fieldwork was conducted, a region where less than 1 percent of pastureland was fenced, community leaders, not government officials, monitored pastureland use and resolved conflict over grazing rights.[67] Rangeland-use contracts in China in the 1990s were apparently not synonymous with private property or households boundaries.[68] Of course, where "fuzzy boundaries," particularly in relation to summer pastures, were the rule, one might expect periodic conflicts over pasture, yet arbitrators chosen from among herding households were credited with successful mediation.[69]

A somewhat different scenario, one that does involve the role of religion, unfolds when we examine pasture mediation in the Tibetan grasslands of Amdo (Qinghai) in China. According to the geographer Emily T. Yeh, who conducted fieldwork in ethnically Tibetan regions of China in the late 1990s, when disputes over grazing rights occurred between groups from different encampments in her area of research in Qinghai, elders from other encampments or lamas were called upon to mediate; the conflicting parties were often required to take an oath "in front of the gods." In other words, recourse to the spiritual world was an integral component of adjudication.[70]

Yeh's article refers to "[t]he government's lack of moral authority to settle disputes, and its reluctance to step in and try," but even more surprisingly, Yeh points to the government's turning to Buddhist lamas to help in resolving pastureland disputes.[71] The tenth Panchen Lama in the 1980s played an important role in directly mediating and delegating authority to lamas to mediate grassland disputes.[72] Most interesting is the fact that, according to Yeh, "many Tibetan officials refer pasture disputes to lamas partly because they too invest them with moral authority."[73]

Yeh's observations on the involvement of Buddhist lamas in adjudicating pastureland conflicts in ethnically Tibetan regions of Qinghai were not unique. A Mongolian scholar from Inner Mongolia who conducted fieldwork in ethnically Mongolian and Tibetan areas of Qinghai from 1996 to 2003 reported "chronic pasture conflicts" that also entailed an ethnic dimension of Tibetan versus Mongolian.[74]

When a particular pastureland conflict was too difficult for the local authorities to resolve, the Tibetans and Mongols involved turned to two "Living Buddhas" of a respected monastery to mediate. Had both sides accepted the Living Buddhas' mediation report, government authorities would then have invested it with legal standing. Even though one side in the conflict ended up rejecting the Living Buddhas' resolution, it is clear that Buddhist involvement in local pastureland issues was the norm, and that all involved acknowledged the respected status of the Buddhist ecclesiasts.

Given the twentieth-century and continuing history of repression of religions in China, including but not restricted to Buddhism, it is intriguing that the Buddhist establishment still plays a significant role in pastoral regions of Qinghai that include both Tibetan and Mongolian populations. This is particularly striking when we examine the limited social involvement of Buddhism in the Mongolian countryside.

While Buddhism has clearly enjoyed a revival in postsocialist Mongolia, the nature of that revival is subject to much debate. Some of the evidence for lack of social involvement is based on a silence in the sources. Ole Bruun's recent fieldwork in Arkhangai Aimag in west-central Mongolia makes no mention of Buddhist lamas undertaking any sort of conflict resolution tasks among herders, which presumably would have been highlighted in the subsection entitled "Interaction between Lamas and Herders" in his chapter entitled "The Politics of Buddhist Revival."[75] In Bruun's description, the postsocialist Mongolian state as of 2001 had not stepped in to oversee pastureland usage and, furthermore, he argues that no "customary" institutions ever existed among herders to allocate or manage pasturelands.[76] Given such an institutional vacuum, one might expect to see a reviving Buddhist clergy exert its moral and practical authority in the countryside. Recent field research in the Khangai Mountain zone of Mongolia attests to herders asking monks to read Buddhist texts to help multiply livestock numbers or to cure illness in a family.[77] None of the anthropological literature on contemporary Mongolia that I have consulted mentions lamas actively adjudicating pastureland disputes. There seems to be a lack of evidence pointing to Buddhist lamas taking active social roles in the countryside.

If Buddhist dignitaries in ethnically Tibetan and Mongolian regions of China can play such a useful role (useful both in terms of the herders' needs and in terms of the lamas' own prestige-maintenance), why do we not see the emergence of Buddhist leadership in the Mongolian countryside? This is a complex question that will necessitate some analysis of Buddhism's historical role in Mongolia as well as analysis of what Mongols nowadays expect of the religion. Recent research into Buddhism's role in twentieth-century Mongolian history may provide some direction. A Mongolian sociologist, Narantuya Danzan, has outlined Mongolian Buddhist lamas' traditional services to rural inhabitants: lamas visited herders' households to attend to matters related to illness, death, and other family issues. While Narantuya refers to herders asking for lamas' assistance with selecting "days and time for movement, settlement, and activities related to animal husbandry," she omits any reference to the role of mediator in conflict situations.[78] Thus, involvement in pastureland conflict resolution was not the norm for Buddhist lamas in traditional Mongolia.

The purpose of Narantuya's research was, in her own words, to "investigate how and why such a devoutly religious people as the Mongols, who had been Buddhists for over 400 years, could so widely discard Buddhism or convert to other religions within the last seven decades."[79] Since 1990, certain factors that have limited the moral authority of the Buddhist establishment along with the growing competition with other religious establishments—particularly evangelical Christian denominations—have further weakened the role of Buddhist lamas in the wider Mongolian society.

A brief review of Buddhism's historical appeal and its displacement through repression may be appropriate here. In terms of rituals and practice, Buddhism in Mongolia had gradually over the centuries accommodated the preexisting beliefs of the pastoral nomadic population—beliefs in a plethora of gods associated with nature. Absorbing elements of shamanist rituals, Buddhism also incorporated a spiritual reverence for the land that supported the herders' way of life. Perhaps the most obvious example of the adoption of a preexisting ritual by Mongolian Buddhism was the spirit-worship at *ovoo* shrines. Ovoo are generally (but not always) rock cairns situated at high mountain passes or at the very summit of a mountain.

To ensure safe passage, a traveler should circumambulate the ovoo three times clockwise and add a stone or other offering to the cairn.

By the early twentieth century, Western travelers would mistakenly identify ovoo-worship as a purely Buddhist phenomenon, so thoroughly had lamas incorporated ovoo-worship into Buddhist ritual practices. The photographer J. B. Shackelford, a member of the 1928 Roy Chapman Andrews expedition to the Gobi, described an enormous ovoo in Inner Mongolia: "The Mongols have a custom of building stone-piles called obos, each stone of which is said to represent a prayer to Buddha."[80] The assumption that ovoo-worship was Buddhist in origin was entirely understandable: Walther Heissig in his study of religions in Mongolia found that all surviving prayers related to ovoo have Lamaist content.[81]

Heissig described in depth the Lamaist suppression of shamanism in Mongolia beginning from the late sixteenth century when Buddhism was reintroduced to the Mongols via Tibet. As Heissig pointed out, shamanism survived as did various gods of nature by allowing their incorporation into the Lamaist pantheon and ritual practices. As Heissig wrote, "Flexible in its defence, shamanism began to camouflage itself through accepting Buddhist phrases and through invoking Buddhist divinities conjointly, until in the course of time it appeared acceptable to the Lamaist church."[82] Conversely, because the early Buddhist missionaries could not snuff out all the gods and spirits that were believed to protect people and their herds from illness, catastrophic weather, and other misfortunes, those preexisting deities were given Buddhist identities.

Even with Buddhism's traditionally close identification with the nomadic population's spiritual attachment to the land, more than 60 years of intense, coerced secularization of values and beliefs in the socialist era (1921–1990) made the so-called revival of Buddhism far less pervasive than some might have predicted. After all, in the 1930s, the government, as directed by the Mongolian People's Revolutionary Party, destroyed monasteries, arrested and often executed Buddhist monks, and thoroughly decimated the Buddhist establishment. Yet, in spite of the apparent diminishment of the Buddhist religion in Mongolia by the end of the 1940s, clandestine worship along with the hiding away of Buddhist texts and ritual objects by many lay believers kept Buddhist knowledge alive even in the worst of times.

In fact, based on recent interviews of Mongolian Buddhist scholars and lamas, Vesna Wallace argues that "during the communist period, Mongolian people in general knew considerably more about Buddhist doctrinal tenets and practices than contemporary Mongols do."[83] As Wallace details in her fascinating research, state support of Buddhism along with many Mongols' tendency to equate national identity with being Buddhist have made the Buddhist establishment quite visible in terms of the symbols of public life. Yet, what about Buddhism's role in the greater society, for instance, at the local level, particularly in the countryside? It is not at all clear that today's Buddhist lamas have the luxury of focusing on the nation's pressing social and economic issues; instead, rural lamas in Mongolia nowadays must focus on the very survival of the religion itself as older lamas pass away and financing of rural temples is meager.

The increasing "Tibetanization" of Mongolian Buddhist rituals may be contributing to a distancing of the Buddhist establishment from lay believers, including rural households. Narantuya's interviewees complained about the Tibetan content of Mongolian Buddhist services; in the words of one informant who identified himself as a believer in Buddhism,

> A number of monasteries and temples have been built. But the teaching of the fourteenth and sixteenth centuries has not been adjusted. Ranges of inexplicable books are read in Tibetan. But people do not understand what has been read and what it brings them.[84]

In other words, in spite of the attempt to rebuild temples and monasteries throughout Mongolia, the social roles of Buddhist lamas remain self-defined in rather narrow terms. Both Vesna Wallace and Narantuya Danzan have suggested that the Buddhist establishment's failure to tackle Mongolia's social problems has allowed Christian missionaries to fill the void. "We give, Buddhism takes" encapsulates the Christian missionaries' own comparison of their outreach services to Buddhist lamas' apparent lack thereof.[85] Narantuya goes even further to argue that since Buddhism was revived in an era of rampant free-market capitalism, Mongols' expectations of deriving tangible as well as spiritual benefits from making monetary contributions to the Buddhist establishment have not been met.[86] In terms of the particular issue of pastureland conflict resolution, however, it is rather obvious that Christian missionaries will not be able to take up the slack left by rural Buddhist lamas; most Christian missionaries are stationed in the capital and have their hands full working on the pressing social and economic needs of Ulaanbaatar's poorer residents.

One has to be careful, however, not to overestimate what contemporary Mongols expect of Buddhism. The October 2008 "Politbarometer," a public opinion survey carried out periodically by the Sant Maral Foundation in Mongolia, asked how strong the role of religion should be in one's life; in one's education; and in politics. The responses indicated a moderate to tepid interest in religion's role overall, with only 21.1 percent opting for the combined categories of "very strong" and "strong" for the role of religion in one's life.[87] The October 2010 "Politbarometer" recorded a slight dip, with only 20.1 percent agreeing that religion should play a "very strong" or "strong" role in one's life.[88] For most Mongols religion is still equated with Buddhism, in spite of the recent inroads made by various Christian denominations. One may speculate with some certainty what the response of Tibetans would be to such a line of questioning, if those polled could be assured of no negative consequences by Chinese authorities. To understand why Buddhism in Tibet, while suffering constant state-sponsored repression, plays a greater social role than Buddhism in Mongolia, which is a multiparty democracy with freedom of speech, we may turn to the important research undertaken over the past decade by two Hungarian scholars, Zsuzsa Majer and Krisztina Teleki.

The two scholars write that in comparison with the earlier, presocialist situation of Buddhism in Mongolia, "today there is not much evidence of the vivid religious life of the past."[89] Despite the fact that many Buddhist temples were established after the 1990 collapse of socialism, very few of those now operate—in fact, about half of all such temples founded after 1990 have closed.[90] My own anecdotal evidence supports this data: in July 2010, while traveling in western Dundgov' Aimag, I came across one such monastery in the sum center of Erdeni-Dalai. Dalai Monastery's outer fence was locked and the grounds and temples were inaccessible with nobody available to unlock the premises in midday. This monastery had been reopened in 1991 after years of neglect. In June 1996, I had been able to tour the monastery grounds at will, but in summer 2010, it was apparently not in active use. In early August 2011, while in Khovd Sum center in Uvs Aimag, I stopped to see the sum's Buddhist monastery, rebuilt in 2006 on the site of the one destroyed in the 1930s. The monastery was locked up for the summer. I was told that the monastery's manager was off in the countryside herding livestock and would return in late August. These two examples of monasteries that were closed up for the summer season suggest a less than vibrant involvement in local communities, perhaps because of staffing insufficiency.

Because those Buddhist monks who survived the purges of the 1930s are now quite elderly or have already passed away, recruiting new monks in the countryside has been extremely difficult. The physical structures were thoroughly demolished in the 1930s, so very few residential facilities exist for monks in rural areas. As Majer notes, when new temples are established nowadays in the Mongolian countryside, they tend to be situated in an aimag center, not in more rural areas.[91] This tendency directly contributes to Buddhism's lack of a social role in the lives of herders, particularly where mediation over pasture usage would entail deeper roots in the local community.

Younger monks in the countryside often feel the pull of Ulaanbaatar where they can receive proper Buddhist training at Gandan Monastery; some choose not to return to the countryside. Monks from the countryside also may end up returning to secular life in order to find employment. Thus, Teleki's characterization of rural Buddhist monks as "not very eminent" is entirely understandable. Because of their lack of social standing, rural lamas are not likely to play a role in pastureland conflict or in other areas of importance to Mongolia's herders.

The numbers also suggest why lamas' social roles are so circumscribed: in all of Mongolia in 2009, there were between 2,000 and 3,000 lamas, with perhaps 1,000 of them living in Ulaanbaatar.[92] Because monastic celibacy is not widely observed in Buddhism as it is practiced in Mongolia, and because of the lack of residential facilities for monks in the countryside in particular, many monks live with their families. The blurring of lines between religious and secular identities may in the long run prove useful to Buddhism's survival and growth in Mongolia, but in terms of social status in today's Mongolia, this blurring of identities would seem to water down the authority of the monk. Majer observes that "it is not surprising that young lamas, having changed their lama robes for lay clothing after finishing 'work' in the monastery in the afternoon, are involved in exactly the same activities as other Mongolian teenagers or adolescents."[93]

When we ponder the ups and downs of the Buddhist establishment in Mongolian history over the past century or so, it is perhaps understandable that the revitalization of the religion since 1990 has not meant a widening social role for lamas many of whom have had to persevere financially and to function with inadequate religious training. As for the narrower question of whether lamas ever adjudicate land use issues, there seems little probability that they will in the foreseeable future become actively involved in this dimension of rural life. Clearly, the moral authority wielded by Tibetan lamas in ethnically Tibetan regions of China has no parallel among lamas in Mongolia.

Secular Authority and Land Use Issues in Post-1990 Mongolia

Secular authority has long trumped religious authority in the Mongolian countryside. During the socialist period, the negdel was responsible for enacting land use policies that were largely drawn up by the government in a top-down hierarchical pattern of decision making that is typical in one-party socialist states. In his 1989 autobiography, *Bounty from the Sheep*, the herder Namkhainyambuu bemoaned "the gap between life and theory":

> Today there are a lot of representatives who don't know about herding but they teach herdsmen, and they grandly announce plans of work to be done within a definite period of time. Is it necessary that they impress us? What are they doing it for? The herdsmen know what to do without their lessons.[94]

In spite of Namkhainyambuu's aversion to meddling by government and party officials, there was certainly no confusion over who was in charge of land and livestock issues in socialist Mongolia. When the negdels ceased to operate as regulatory institutions in the early 1990s, however, a vacuum opened up in the countryside waiting to be filled by "traditional" or "customary" institutions that were, at best, defined rather vaguely by most observers. In recent years, the sum and bag levels of local government have been deemed legally responsible for the adjudication of land conflicts. Yet, as recent scholarship has pointed out, the local administrative units often work in tandem with "experienced elders," who are themselves leaders of khot ails.[95] Thus the local representatives of state authority, rather than dictating policy, cooperate where possible with the most respected members of herding communities.

Finally, the pending legislation on pastureland that was under discussion in the Mongolian parliament in 2011, if enacted, would delineate with unprecedented clarity the procedures for adjudicating pasture disputes.[96] Herders' self-governing organizations and the local, bag-level *khural* (assemblies) would bring the aggrieved parties together. If unsuccessful in resolving the issues at this level of local society and government, the governor "of a higher level," for example, the sum, and then the aimag, would attempt to settle the dispute. If none of these authorities could resolve the contested pastureland issue, a judge would then be the ultimate arbiter. It remains to be seen whether this legalistic formula can be grafted onto the preexisting cultural norms of Mongolian rural society, but the proper combination of bottom-up and top-down authority will be the key to success.

Winter Livestock Enclosures: What They Tell Us about Land Use

For the pastoral nomadic household, the key area in which land legislation and local practice intersect is the winter encampment with its livestock enclosure and winter pastureland. Because winter and spring in Mongolia see the harshest weather, it is crucial for herders to lay claim to winter grazing lands that offer some natural protection from the elements (often mountains that block winds) and offer access to water; it is also crucial to build livestock pens or enclosures in the protected lee of mountains where possible. In fact, the authors of the Open Society's 2004 survey of herder opinion concluded that the insufficiency of livestock enclosures at winter and spring campsites was an even greater problem than insufficiency of water sources and pasture (which the herders themselves viewed as more pressing).[97]

The fieldwork of anthropologists and geographers offers us a window into this sliver of rural life in Mongolia. In particular, the fieldwork of

Mari Kazato in Telmen Sum, Zavkhan Aimag in October 2003 and July-September 2004 offers excellent snapshots of land use in practice. This section will also incorporate my own firsthand observations from travel in June-July 2005 in Zavkhan and Khovd Aimags; August 2006 in Ömnögov' and Övörkhangai Aimags; June-July 2010 in Dundgov', Ömnögov', and Selenge Aimags; and July-August 2011 in Uvs Aimag. The accompanying photos of winter enclosures show both variety in size and building materials as well as consistency in siting the structures to protect the smaller livestock (sheep, goats) from winter winds (see figures 5.1, 5.2, and 5.3). For instance, in the absence of trees to use as poles for the traditional lean-to in forested areas, herders in Khovd Aimag rely on rocks to create livestock shelters. Whether constructed from wood or rock, livestock shelters usually have dried livestock dung to seal the walls and a floor of dried sheep and goat dung.

As Mari Kazato argued in 2006, winter campsites—the pastures and livestock shelters associated with them—"lie at the frontier of legal land rights in pastoral regions."[98] In other words, while state ownership of those pastures associated with winter campsites as of 2006 remained uncontested and based in constitutional law, local community-based customs and households' maintenance of such winter campsites determined actual usage rights. By 2010–2011, however, in at least three aimags, namely Ömnögov', Selenge, and Uvs Aimags,

Figure 5.1 Winter livestock enclosure, Zavkhan Aimag, summer 2005.

Figure 5.2 A herder does summer repair work on a winter livestock shelter, South Gobi Aimag, 2006.

Figure 5.3 Winter livestock shelter near Achit Nuur, Uvs Aimag, summer 2011.

state ownership was modified by long-term leases vis-à-vis winter livestock enclosures and surrounding pastures as we shall see below.

As others have noted, summer pasture encampments may have livestock pens or shelters, but these are of a far less permanent nature than the well-maintained livestock shelters adjacent to winter pastures.[99] Mongolian scholars have noted the historical tendency for winter pastures to be more clearly demarcated than other seasonal pastures. Writing about the Autonomous Period (1911–1919) in Mongolia, the historian B. Shirendyb noted that in the "northern parts of the country the process of delimiting winter pastures even for individual households was continuing (although this was not possible in Gobi areas)" and that pasture boundaries were "becoming ever more closely defined."[100] Shirendyb's parenthetical comment vis-à-vis the Gobi being an exception to patterns found in other parts of Mongolia is a theme that we touched upon in chapter 2.

Summer pastures and often autumn pastures have been described as "open-access resources"[101] because of the more dependable access to water and forage as compared with winter pastures. Mari Kazato found in her fieldwork region in 2003–2004 that spring and autumn corrals (*khavarjaa* and *namarjaa* respectively) that had been constructed in the socialist era (1921–1990) had been dismantled and the materials were then used for other projects "because herders did not consider these structures necessary."[102] Thus, it is understandable that winter campsites, crucial to the survival of one's herds, have been subject to governmental legislation that promotes some degree of privatization.

With the dismantling of collectives by the end of 1992, state authority over pasture usage was virtually nonexistent. The distribution of the assets of the former negdels led to a variety of outcomes well detailed in the case studies of Fernandez-Gimenez and Kazato.[103] As Kazato emphasizes, households that had been defined by their membership in a socialist collective tended to adhere as a unit even in the absence of common kinship. Former negdel administrators in some cases profited from the distribution of collective assets including the winter campsite livestock shelters. Herders who wished to reclaim winter campsites and structures that predated the socialist era of negdels were apparently not successful in pressing their claims. The problems and disputes surrounding distribution of winter campsites mirrored the problems of distribution of other collectively held assets throughout Mongolia in the early 1990s.

Mongolia's parliament debated new land legislation in the early 1990s, with some members of parliament favoring movement toward land registration and titling. Foreign developmental economists and advisers to the Mongolian government often assumed that privatization of land was essential to the success of a market economy and that land privatization would benefit the natural environment.[104] The 1994 Land Law did not go so far as to authorize private ownership of land, but the Land Payments Law (also 1994) did allow the state to lease campsites and pastures to individuals. The 1994 legislation was ambiguous on the hierarchy of authority involved in adjudicating conflicting claims over pasture and campsites, although the onus seemed to fall on local officials.[105]

It was not until 1998 that the government actually issued ezemshigchiin gerchilgee (certificates of possession) for winter campsites based upon reasonable evidence that a herder had continuously occupied and maintained a campsite livestock shelter. As Fernandez-Gimenez has pointed out, the allocation of these certificates was potentially flawed in cases where a group of households collectively used a campsite: certificates of possession usually listed only one name, that of the senior herder.[106] In Kazato's region of fieldwork (Telmen Sum, Arkhangai Aimag), revised certificates of possession were distributed in 1999; these certificates authorized the leasing for a 60-year period of winter campsites to herding households.[107]

In June-July 2005 along a route from Telmen Sum in Zavkhan Aimag to Manhan Sum in Khovd Aimag, I chatted with several herding households regarding their opinions of the certificates of possession for winter livestock enclosures and the associated winter pasturelands. Virtually all the herders with whom I visited that summer did have the certificates of possession and most had been granted for a period of 40–60 years as a sort of long-term lease. While no payment had been necessary to obtain the certificates, the herders paid a yearly fee (or tax). A herder in Telmen Sum showed me his ezemshigchiin gerchilgee with state seal affixed: the certificate stated that his family had possession of their winter enclosure and winter pastures for 60 years as long as they continued to pay a yearly fee. Only winter pastures were subject to certificates of possession; summer pastures continued to be state owned and constitutionally guaranteed open to public access.

I gained the impression that most herders favored the continued use of certificates of possession. Only one herder with whom I spoke in a mountainous summer pasture area some 25 kilometers southwest of Khovd City in Khovd Aimag pointed out the negative consequences of the certificates: he claimed that the summer pastures were much more limited in scope now that the other side of the nearby mountain had been claimed as winter pasture by another herder. In fact, the 2002 revised Land Law clearly prohibits pasture around winter enclosures from being used in other seasons.[108] Thus, the herder on the other side of the mountain was clearly within his rights to deny other herders summer usage of the pasture in close proximity to his winter livestock enclosure. The herder whom I interviewed, however, claimed that his sheep and goats were not fattening up properly because of such limits on pasture usage in the summer months. Overall, however, the certificates of possession were seen as a positive step in the right direction among this small sampling of herders in Zavkhan and Khovd Aimags in 2005.

Variation rather than uniformity was the rule across Mongolia in the years 2005 and 2006 vis-à-vis the certificates of possession. A camel-herder in Ömnögov' Aimag near Bayanzag (the "flaming cliffs" of Roy Chapman Andrews) told me in August 2006 that herders in the South Gobi did not use the ezemshigchiin gerchilgee because of the low population-to-land ratio, and because, in his opinion, local herders simply knew whose grazing land was whose. Pressed to explain further, the camel-herder added that local herders used ovoo to mark boundaries between pastures. When I mentioned that herders in Zavkhan and Khovd *did* use certificates of possession, the camel-herder ascribed this practice to the number of herders in those aimags (in other words, a perceived herder-to-land imbalance).

Even within Ömnögov' Aimag, there were wide variations in the use and terms of use of winter enclosures. For instance, a herding family with 60 camels and some horses, goats, and sheep just north of Bulgan in Ömnögov' Aimag did have a certificate of possession for winter/spring pastures and a livestock enclosure. The certificate was good for three years, and they paid 7,000 *tögrög* a year as a fee. The family was temporarily camped at their winter quarters en route to moving north in early August.

The lack of standardization in the use and terms of use of certificates of possession that I encountered in 2005 and 2006 across Mongolia perhaps stemmed from the activism or, conversely, a lack of involvement of local sum governors from locale to locale. Legislation emanating from the parliament in Ulaanbaatar might also have been interpreted differently according to regional and local realities and priorities. By 2010, however, the picture had changed and suggested a greater acceptance of actual ownership of winter livestock enclosures and their associated winter pastures.

In July 2010, a camel-herder near Khongoryn Els in Ömnögov' Aimag explained to me that summer pastures were indeed still open access, though he and his family tended to come back to the same summer grazing land every summer. This herder's winter enclosure and winter pastures were about 6 kilometers away from his summer encampment. All herders in Sevrei Sum (county), he continued, had registered their winter enclosures and winter pastures with the sum government office. Each winter enclosure was supposed to be 5–6 kilometers away from the adjoining ones, so as to prevent conflict over scarce winter resources. The sum government office had a map of all winter enclosures and associated grazing lands. Most interesting was the assertion that there was no time limit on the certificates of possession; in other words, they were not leases. In fact, the camel-herder claimed that he would be able to pass his winter livestock enclosure and pastures down to the next generation and beyond. This was the first real evidence of private land ownership that I had ever encountered in the Mongolian countryside.

A similar situation vis-à-vis ownership of winter livestock enclosures exists in Selenge Aimag in northern Mongolia. Not far from the restored eighteenth-century Amarbayasgalant Monastery in central Selenge Aimag, I visited with a 74-year-old herder, father of 12 and grandfather of "too many to count" in his words. This herder, who owned about 1,000 head of livestock, insisted that everyone he knew owns their own winter enclosure and that there is no certificate of possession system in Baruun-Büren Sum. No paper documentation certified his family's "ownership" of their winter quarters. His winter livestock enclosure was about 7 kilometers from his summer pastures, and his household moved by truck (which, as a fairly well-to-do herder, he owned). He even laughed with amusement when I inquired whether he used yak or khainag-drawn carts to move between seasonal pastures.

When asked whether life was better for him under socialism in the negdel system or now in the free-market system, the elder herder replied that life was better now. He added his opinion that ultimately it did not matter what type of government was in power, since herders would shape their own lives and destinies. When I posed the same question to an older herder in Uvs Aimag in summer 2011, this grandfather-herder opined that life was better under the negdel system when there was economic equality. Yet, nowadays, he added, one could certainly prosper—if one worked hard.

In fact, this feeling that one is ultimately responsible for one's own fate is quite typical among herders, as the October 2010 Sant Maral "Politbarometer" Survey suggests: people in the countryside overwhelming chose the answer "Rather on myself" in response to the question "What do you think, does your future depend on your own achievements, the State, or on other forces (like churches, trade unions, firms, the press, TV/radio, communities, business relations), that could influence the economy?"[109] According to the survey, 48.6 percent in the countryside believed in self-determination of one's future, with the state garnering second place with 32.3 percent. This bespeaks a certain duality between herder self-confidence in determining outcomes if one exerts oneself diligently and the expectation that government should step in to help when needed. The continuing belief that government should be activist is likely a legacy of the socialist era.

My most recent travel in the Mongolian countryside involved an extended itinerary through Uvs Aimag in Mongolia's northwest, where I found the use of certificates of possession to be widespread. All the herders with whom I spoke replied positively to my question about whether they had the certificates for winter pasture and enclosures. Some even had certificates for spring pastures and livestock enclosures. There was, however, variety in the leases' duration: some certificates were issued initially for 10 years, then extended to 30 years, and, most commonly 50 years. A yearly tax on a certificate of possession amounted to about 5,000 tögrög, a negligible amount. A few households had certificates of possession drawn up for 40 years; a few had 15-year certificates that once they expired could be extended to a term of 30 or 40 years. In one case, in a ger near Shaazgai Nuur in Khovd Sum in Uvs Aimag, I was shown three certificates of possession, two for 30-year leases of winter pasture and one for a 15-year lease, reflecting different arrangements for three households related by kin. According to one of the herders in this ger, herders may also obtain certificates of possession for spring pastures with the same guarantees and annual tax. I also was told that if a herding household for some reason decided not to use its winter or spring pasture, the leasee may rent it out to earn money.

Most interesting was the case of an older herdswoman, a grandmother, in the same sum who apologized that she could not show her certificate of possession because she had handed it over to the bank as collateral for a bank loan that she had taken out. If she were to default on the loan, the bank would become the owner of the certificate and could sell it to another household. Incidentally, the grandmother-herder had boxes of her own handcrafted palm-sized felt animals, a selection of mice, camels, and snow leopards (all found in that part of Uvs Aimag) for sale to supplement her income.

All in all, the certificates of possession in Uvs Aimag were in wide use, and, in spite of variation in the length of time of a lease, the documents reflected a certain degree of standardization. The notion of mortgaging a certificate and the idea of renting out one's unused winter pasture and livestock shelter suggest an evolving fiscal dimension in land use in rural Mongolia. Winter livestock enclosures will continue to reflect the evolution in conceptualization of land rights among pastoral nomadic households in Mongolia.

Water rights do not seem to command the same degree of government attention and involvement in Mongolia. I asked a few questions at the Uvs Aimag Land Office in Ulaangom of an aimag-level water specialist whose duties included monitoring aimag-wide wells for functionality and purity. Herder families in Uvs Aimag have the right to possess and use their own wells, but most herder households share wells with other households. While it is possible to legally register one's well, in reality very few herders bother to do so in Uvs Aimag. Thus, there is nothing analogous in the realm of water rights to winter pastureland certificates of possession.

While my random conversations with herders obviously cannot pretend to reflect the views of all Mongolian herders, they do at least reflect a variety of opinions from several aimags. I leave it to others to produce more thorough statistical surveys of herder opinions.

THE DIFFICULTY OF GAUGING HERDER OPINION ON LAND OWNERSHIP, PRIVATIZATION, AND FENCING

Overall, inherently contradictory impulses have been at work in the land legislation of the postsocialist period. The commitment to maintaining access to pastureland—a commitment that is explicit in both the 1992 constitution and the 1994 Land Law—coexists uneasily with the emerging approach to land as a marketable commodity, as reflected in the 2002 Land Law. The evolution of the certificate of possession system underscores the ambiguities and inconsistencies of land tenure in rural Mongolia in this era. In 2004, David Sneath aptly described the peculiarity of "a property regime in which pasture land [is] neither open-access public property nor in private ownership."[110]

Gauging herder opinion on privatization of land has been difficult in the postsocialist era, no doubt owing to the fact that opinions may differ by region and may fluctuate in response to particular challenges posed by climate, proximity to mining, tourism, and other competing forms of land use. When the 2002 Land Law was approved by the Mongolian parliament on June 28 of that year, the Mongolian press reported that the Law passed "despite intense opposition from not only political parties but from the population as well."[111] A broadly worded question regarding land privatization was included in both the October 2002 and the March 2003 Sant Maral "Politbarometer" surveys: "Was the decision to privatise land right or wrong?" Only 43 percent of respondents in the countryside in October 2002 approved the decision, but the rate of approval in the countryside rose to 49 percent in the March 2003 survey.[112] A more incisive, though geographically confined, polling of opinion among herders in Mongolia regarding land use issues was undertaken by Maria E. Fernandez-Gimenez and B. Batbuyan. In their 1994–1995 study, which was followed up by a 1999 survey of households in Bayankhongor Aimag, the two researchers found that "[h]erders continue universally to oppose privatization of pasture in any form, and perceive possession contracts over pasture as a form of privatization."[113] Many of the herders interviewed by Fernandez-Gimenez and Batbuyan felt that transforming pastureland into private property would create, rather than resolve, conflict. One of course would ideally like to know whether the herders in Bayankhongor Aimag would feel the same in 2011 as they did more than a decade earlier.

Fernandez-Gimenez and Batbuyan along with other researchers, including Robin Mearns, tend to reject the applicability of property ownership in the pastoral nomadic context of rural Mongolia. The research conclusions of Fernandez-Gimenez and Batbuyan are encapsulated in the phrase "the strong ethic of reciprocal access," which they also refer to as "a moral economy of the steppes."[114] In similar tones, Robin Mearns has referred to the "deeply rooted ethic of open access among Mongolian herders" that, he believes, would undermine any rigid application of grazing land access regulations.[115] As we have seen in earlier chapters, Mongolia's herders do take account of the variability of climate and pasture quality in deciding upon seasonal migration routes; they are certainly still a long way away from adhering to strict scheduling or strict property guidelines for pasturing their herds much of the year. The real changes have been limited to winter livestock enclosures and the surrounding pastures.

In peri-urban zones, the pasturelands abutting the ever-expanding city of Ulaanbaatar have witnessed a noticeable increase in fencing. This trend is not surprising given the mixed use of such lands that may host dachas built by Ulaanbaatar's new elite, hotels and ger camps, semi-intensive dairy farms, as well as some land for livestock grazing. In the countryside, however, fencing of pasture is still very unusual. One can, of course, find stray examples in press accounts, as with the case of the Bayankhongor Aimag herder, who in 2004, complained of the lack of "legal protection to recognize ownership of pastureland that has been fenced off."[116] As reported in the press account, this particular herder had fenced pastureland for the purpose of grazing his livestock, haying, and planting vegetables; he had also planted trees to stop erosion. He felt, however, that the lack of land ownership documentation could lead to his eventual loss of use of the enclosed pastureland if another herder were to make a claim of his own.

Overall, there does not seem to be a general consensus among Mongolian herders that fencing would be a good management option; the traditional use of ovoo to demarcate pastureland suffices. Given the dire consequences of extensive fencing of pastureland in Inner Mongolia in China, it is hard to make the argument that pasture enclosure will aid sustainable use of Mongolia's pasturelands. Certainly, the fieldwork conducted by Dee Mack Williams in Inner Mongolia suggests that a government-driven pro-enclosure policy can lead to an unintended extreme stratification in wealth among herders and can impact negatively upon pastureland sustainability.[117]

The traditional Mongolian hospitality that is rooted in practicality when circumstances demand such, has not been overly affected, at least so far, by the privatizing of winter livestock campsites. Fernandez-Gimenez and Batbuyan have referred to "the moral economy of reciprocal relations that underlies the nomadic way of life" in Mongolia. Potentially more disruptive of traditional Mongolian concepts of hospitality and reciprocity is the intrusion of mining claims and all the commercial operations related to mining in many areas of Mongolia. Commercialization associated with large-scale mining in some areas of the country will undoubtedly affect aspects of the pastoral nomadic way of life, while at the same time creating new jobs that may augment the herder economy. In fact, the mining sector is likely to produce jobs for unemployed and underemployed residents of sum centers as well as jobs for herders. The topic of mining is a focus of chapter 6.

Placing Mongolia in Perspective: Herders and Pasture in Kazakhstan

The challenges faced by Mongolia's herders are not unique in Central Eurasia. A brief look at Kazakhstan's pastoral nomadic herders past and present will reveal both similarities with, and differences from, the Mongolian case in land use, access to pastureland, and interaction with governmental authority.

A gripping memoir by the Kazakh nomad-turned-teacher Mukhamet Shayakhmetov provides us with a wealth of information about pastoral life in precollectivization Kazakhstan in the late 1920s to early 1930s. Recalling moves of 150–200 kilometers to summer pastures in eastern Kazakhstan, Shayakhmetov's recollections are certainly tinged with nostalgia for a way of life that came to a sudden, disastrous end: "Each move was like a festival, especially for us children: everyone was happy, and dressed up for the occasion."[118] Writing many decades later, Shayakhmetov no doubt romanticized the migration cycle of the Kazakh herder: "These moves were easy for us, as they had been developed to a fine art."[119] Who determined the rights of access in those days? Shayakhmetov writes simply that "we were always on the move between pastures, following the routes established by our forefathers."[120]

Such vagueness about rights of access to natural resources (pastureland, water) is typical not only of traditional Kazakh nomadism but also of most nomadic groups in Central Eurasia. The flexibility of such arrangements generally went hand-in-hand with the absence of centralized political-administrative control. The obvious exceptions to this state of fragmentation were periods of nomadic military expansion, such as under the Xiongnu nomadic confederation (roughly third century BCE to third century CE) and the Mongolian empire of the thirteenth and fourteenth centuries. The Kazakhs, however, never formed such a large-scale, centralized, pillaging and redistribution regime, one that allowed a khan to build loyalties through redistribution of loot from sedentary societies. As the authors of a recent piece entitled "Ideology, Land Tenure and Livestock Mobility in Kazakhstan" note, the natural environment contributed to fragmentation in Kazakh society: "The Kazakhs occupied a mountain-steppe-desert environment in which pastoral resources were widely dispersed and erratically available. The dispersal of natural resources militated against the creation of a centralized political system."[121]

During the centuries that the Kazakhs came under Tsarist Russian control, pastoral nomadic migration routes and the pastures themselves were increasingly encroached upon and appropriated. Nineteenth-century Russian legislation redefined all Kazakh lands as Tsarist Russian state lands. Migration by Russian peasant-farmers into traditionally Kazakh lands was spurred by the emancipation of Russian serfs in 1861 as well as by the growing need for farmland. This was parallel in scope to Chinese migration into Inner Mongolia whose pasturelands were expropriated by land-hungry Han Chinese migrants from interior provinces of Qing China in roughly the same period. The Russians irreversibly altered the Kazakh landscape from one of steppe pastures to one of farm fields. Kazakh herds shrank in size as pastureland and migration routes became inaccessible, and many Kazakhs took up farming often in combination with livestock herding. By 1919, it is estimated that 90 percent of Kazakhs sowed grain to be harvested for winter feed for livestock.[122]

The Kazakh case reminds us that nomadic agriculture historically has differed from sedentary agriculture in Central Eurasia. Thus, Shayakhmetov recalls his family sowing wheat and grain crops at the spring pastures; upon return in the fall, hay would be harvested: "Nomads did not put aside a stock of fodder for winter, except for the possibility of a couple of horses or animals falling sick."[123] The casual, less intense, nature of nomadic agriculture sharply contrasts with intensive sedentary agriculture, a theme that we shall return to in chapter 6.

Alimaev and Behnke have characterized the period from about 1800 to 1929 as witnessing the emergence of a "hybrid land tenure regime combining elements of both Kazakh and Russian legal traditions, common and private property."[124] Recent research suggests that far from being passive victims of Russian colonial expansion, nineteenth-century Kazakh nomads employed various strategies to manipulate Russian laws and regulations to their own benefit.[125] While disputes over pasture and water access had always existed among the Kazakhs as among other pastoral nomadic groups, the latter half of the nineteenth century saw unprecedented pressures on the land as a result of the flood of Russian settlers and the imposition of new administrative borders. According to traditional Kazakh customary law (*adat*), claims over disputed pastureland would often be resolved by acknowledging seniority in social or kinship hierarchies. In addition, land disputes might be forestalled by staking or marking one's pastureland; making oral claims to land; or simply claiming grazing lands by setting up one's encampment in a "first come, first served" manner.

With the introduction of the 1868 Provisional Statute, all Kazakh lands were subdivided into administrative units headed by Russian bureaucratic appointees. The drawing of boundaries with the intent of controlling and taxing Kazakh nomads obviously interfered with traditional migration routes and furthered the ultimate Tsarist Russian goal of settling Kazakh pastoral nomads. In this regard, Soviet policy and Tsarist policy aimed at similar results. While Kazakh elites took advantage of lands granted for agricultural use in return for service to the Tsarist state, the issue of Kazakh ownership of land, particularly pastoral grazing lands, remained unresolved.

As restrictive as Tsarist Russian regulations were, the Soviet era of Kazakh history represented a concentrated assault on all aspects of Kazakh life. Beginning in 1929, the Kazakhs fell victim to Stalin's imposition of collectivization and sedentarization policies. The harrowing account of Mukhamet Shayakhmetov, whose family began to feel the dire effects of these policies only in 1930–1931, attests to the immense loss of life—both human and livestock—primarily from the famine that itself was a direct result of these policies. A conservative estimate counts 1.5 million Kazakh deaths in the 1930s, with about 80 percent of Kazakh herds wiped out in four years (1928–1932).[126] While collectivization failed, the forced settlement of the pastoral nomads had a lasting impact: by the end of 1936, only 150,000 or so Kazakh households practiced traditional herding.[127]

Still, a very limited number of Kazakh livestock breeders continued to nomadize throughout the remainder of the Soviet period. The higher cost of settled livestock husbandry and the availability of unused, more distant pastures contributed to this reversal of the previous state policy of sedentarization of all nomads.[128] Herders' migration routes were, however, drastically curtailed since much of Kazakhstan's pastureland had been converted into grain fields. In many cases, herds remained within the confines of state farms in the Soviet era.[129] Kazakh herders benefited from technological improvements paralleling those occurring in the same period in Mongolia, for instance, modern veterinary care for animals and state-supplied transportation for longer migrations. On the down side, Kazakh herders found that they had little input in deciding upon the timing and length of migrations or the retention of lands as pasture. As the memoir by the Mongolian herder Namkhainyambuu attests, such centralized decision making was typical of socialist state oversight of herding practices in Central Eurasia under Soviet influence.[130]

Decollectivization and privatization in Kazakhstan, as in Mongolia, in the 1990s were riddled with corrupt practices that led to extreme inequalities in the redistribution of collective-farm and state-farm assets. These inequalities affected both farmers and herders alike. Livestock numbers plunged in the decade of the 1990s: the number of sheep, for instance, decreased by 70 percent.[131] The newly introduced land laws of the 1990s allowed the long-term leasing of agricultural land, but these laws did not function well for pastureland use in Kazakhstan. By default, non-leased rangeland became state land. Herders found land registration to be expensive, and the end result of having to pay tax on leased land was unattractive.[132]

Most observers have concluded that since decollectivization began in the early 1990s, the migration patterns of Kazakh herds have shown greatly reduced mobility. This is related to the growing tendency of herders to settle down in one dwelling year round; fewer herders now than in the Soviet period live in yurts in summer pastures.[133] More remote pastures are often vacant, with wells and outbuildings having fallen into disrepair. In the absence of state support to which herders had become accustomed, moves to new pasture could no longer depend upon state-supplied transportation. As a result, there has been a marked tendency among Kazakh pastoralists to graze their flocks near settlements, leading to problems of overgrazing in the vicinity of villages. Presumably, Kazakh herders were and continue to be drawn to settlements because of educational opportunities for their children, economic opportunities such as jobs for family members, and the overall conveniences of proximity to a settlement.

This parallels the rural-to-urban trend seen in Mongolia that occurred a decade later. In the case of Mongolia in the 1990s, herding presented an economic safety net in the aftermath of the postsocialist economic crash: many people fled unemployment and food shortages in Ulaanbaatar to join relatives with herds in the countryside or to attempt to establish themselves as new herders in rural areas. In fact, the number of herders in the 1990s in Mongolia swelled dramatically: less than 18 percent of Mongolia's work force worked in the herding sector in 1989, but by 1998, 50 percent of the national work force worked in herding. In addition, the national herd census showed an increase of over 20 percent between 1990 and 1998, rising from 25.9 head to 31.9 head in that eight-year period.[134] It was not until the late 1990s and beyond that Mongolia saw a reverse surge from the countryside to urban and peri-urban locales. As studies have shown, the desire to access better schools for their children has been a central motivation inducing a significant number of herders to graze their herds as close to aimag and sum centers as possible. This in turn continues to contribute to overgrazing in peri-urban locales.

In Kazakhstan, however, it seems that the economic crisis of the 1990s initiated a more immediate rural-to-urban shift. At an international conference on animal husbandry in Kazakhstan in 1999, participants noted the abandonment of the more distant pasturelands after the early 1990s. Some participants urged a return to nomadic and seminomadic pastoral herding that would make use of these vacant grazing lands. Apparently, by 1999, there was a small-scale resurgence of migration to distant pastures, but this entailed the oldest and youngest family members staying behind in a settlement, while the husband and wife in the family migrated with herds to the remote pastures.[135]

Interestingly, Alimaev and Behnke see the results of the 1930s radical collectivization and the 1990s sudden decollectivization as similar: "Each was followed by an economic depression that depopulated the countryside, destroyed rural infrastructure, and caused a massive crash in livestock numbers."[136] Given the opposing ideologies at work behind both disasters, it would seem that Kazakhstan's herders have been involuntary subjects of large-scale state-sponsored economic experiments that have done little to improve their way of life. Neither socialism nor capitalism, as implemented in rural Kazakhstan, has provided reliable and long-term economic and social support for pastoral nomadism. Alimaev and Behnke credit "the tenacity of pastoral communities" in Kazakhstan for overcoming the socialist and postsocialist pressures put on herders to settle down and abandon migratory patterns.[137]

There is indeed some basis for optimism that cultural tenacity—the desire to continue a way of life in the face of pressures to abandon it—coupled with continuing access to open rangeland that is state owned may allow the remaining Kazakh herders to continue as modern-day pastoral nomads. Yet, the stream of sometimes contradictory land reform laws issued over the past several years, starting with the first Land Code of November 1990, may deter some Kazakh herders from venturing into long-unused rangeland. Certainly, the land laws suggest a governmental indecisiveness over the extent of land privatization in the future.[138] The 1991 constitution of Kazakhstan specifies that only the state may define terms of land ownership; by default all land in Kazakhstan belongs to the state. If Kazakhstan's livestock herds eventually rebound in numbers, regulation of pastures may become a pressing priority especially if herders begin to contest the most desirable grazing lands.

It has to be noted that there is a major difference between herding in Kazakhstan and herding in Mongolia: the overwhelming majority of Kazakhstan's citizens are urban and agricultural, thus creating a very different economic environment in which pastoral nomadic herding represents a distinct minority way of life. Yet, herders in Kazakhstan face challenges similar to their counterparts in Mongolia in terms of potential conflict with mining interests and the ground and water pollution associated with under-regulated mining industries, though in Kazakhstan uranium and thorium are the culprits in contaminating soil and ground water.[139] In Mongolia, the increase in national herd size coupled with competing land use interests (mineral, particularly gold, exploration and exploitation) have pushed the state to face the ongoing need for land legislation. The grasslands and deserts of both Kazakhstan and Mongolia face environmental disasters unless their respective governments step up to the task of creating and implementing strict regulations to protect pasturelands, rivers, lakes, and wells from the negative consequences of mining. If the picture looks a bit more hopeful in Mongolia, that would be because Mongolia's democracy, while young in age, has not yet been so thoroughly infected by corruption as have the "democracies" of the post-Soviet "Stans" (Kazakhstan, Kyrgyzstan, Turkmenistan, etc.).

CHAPTER 6

AGRICULTURE, MINING, TOURISM: COMPETING INTERESTS IN LAND USE

RETHINKING THE ROLE OF AGRICULTURE IN MONGOLIAN HISTORY: OWEN LATTIMORE AND THE QUESTION OF AGRICULTURE

During the 1986–1987 academic year, while I was an assistant professor in the Department of East Asian Languages and Civilizations at Harvard University, I had the pleasure of meeting and chatting with Owen Lattimore (1900–1989), the premiere American expert on Mongolia of his generation, at a few lunchtime talks sponsored by the Committee on Inner Asian and Altaic Studies. On one such occasion, when I was seated next to Lattimore, I asked him whether he thought that agriculture had played any significant role in Mongolian history. His reply was fascinating. Lattimore pointed out that agriculture has been practiced by most nomads in Central Eurasia, but in minimal and very specific ways. The type of agriculture that Owen Lattimore described was essentially nomadic agriculture: rough wooden ploughs broke up the earth; seeds were sown by hand; wheat or other grain crops were left unprotected as nomads pastured elsewhere; and either sickles or bare hands harvested the crops in autumn.[1] Years later, when reading the 1910 report of the Moscow Trade Expedition to Mongolia, I discovered a photograph of a very simple wooden Mongolian plough as well as a photo of Mongols harvesting wheat near Khovd in western Mongolia.[2] This type of nomadic farming was obviously very different from crop farming practiced by Chinese farmers in Mongolia, a fact that was abundantly evident to the Russian merchant-explorers who described the activities of Chinese farmers in the environs of Uliastai in 1910.[3]

Reflecting upon that brief conversation so many years ago, it seems appropriate to review the important contributions that Owen Lattimore made to our understanding of agriculture's role in Mongolian history. Building upon Lattimore's foundation, newer research allows us to sift stereotypes from reality in looking at agriculture in a part of Central Eurasia better known for pastoral nomadism.

What did Owen Lattimore write about agriculture in Mongolia? Two of Lattimore's articles written in the 1930s very nicely encapsulate many of his observations on agriculture: "The Geographical Factor in Mongol History" and "On the Wickedness of Being Nomads." "The Geographical Factor in Mongol History" is the article in which Lattimore wrote the famous line "it is the poor nomad who is the pure nomad."[4] In this article, Lattimore described sporadic agriculture in Mongolia, a prime example being that of the Orkhon Turks who practiced agriculture at the height of their power in the eighth century. Lattimore characterized the intermittent occurrence of agriculture in Mongolia as part of a cycle of flourishing prosperity that then modified nomadic society, leading to the collapse of nomadic leaders' power, and ultimately contributing to the disappearance of agriculture: "the poorest parts of the steppe remained a permanent reservoir in which the essentials of the nomadic way of life were preserved."[5]

Lattimore turned his famous phrase around in his brief discussion of resistance to collectivization in the late 1920s to early 1930s to suggest that nomads could withdraw and survive: he drew the historical lesson that "the poor nomad is the pure nomad, best able to survive under the strictest conditions of the old life, and at the same time best able to evolve into new ways of life."[6] This lesson, which Lattimore was applying to a difficult chapter in Mongolian history (late 1920s–early 1930s), could certainly be applied to a more recent chapter in Mongolian history—the decade of the 1990s, when the economic crisis pushed many people back into the pastoral nomadic way of life as a survival strategy. The anthropologist David Sneath has estimated that the population of nomadic herders more than tripled between the years 1989 and 2001 to constitute almost 50 percent of Mongolia's working population.[7]

As the historian William T. Rowe has pointed out, Owen Lattimore's views on the interrelationship between the natural environment and human society evolved over time; by the 1930s, Lattimore, as Rowe writes, "had begun to pull away from ecological determinism."[8] In fact, in "The Geographical Factor in Mongol History" Lattimore challenged "geographical determinists who believe that primitive human society is controlled rigidly by its environment."[9] As this relates to the topic of agriculture in Mongolian history, Lattimore carefully pointed out the complex interrelationships between forest-dwellers, agricultural communities, and steppe nomads, emphasizing that agriculture had a long, if not continuous, history in the north of present-day Mongolia.[10] In addition to rebutting ecological determinism, Lattimore argued against a sort of evolutionary progression from what was the once accepted sequence of hunting society to pastoral nomadic society to agriculture and then on to modern industrial society, a sequence that is no longer accepted by scholars.[11] When pastoral nomads did switch to agricultural pursuits, Lattimore advised that this "should not however be confused with progress from the primitive to the civilized."[12]

In his article "On the Wickedness of Being Nomads," written in 1935, Lattimore again prodded his readers to rethink categories when he questioned the validity of "an entirely artificial line... drawn between 'civilized' agriculture and [the] 'primitive' pastoral economy, dependent on livestock."[13] Here, Lattimore argued strenuously for the suitability of pastoral nomadism, not agriculture, on the steppe-lands of Inner Mongolia in China. This is a theme to which Lattimore returned in many of his writings, including "The Geographical Factor in Mongol History." Sounding eerily of the present, Lattimore, writing in 1938, stated: "Both the cultivation of marginal areas and the overgrazing of stock in true steppe areas can ruin the soil, create deserts and 'change the climate.'"[14] Lattimore even mused over the possibility (highly unlikely) of Chinese colonists in Inner Mongolia adopting the environmentally better-suited Mongolian way of life:

> Were it not for the operation of economic and political factors (of which railways and modern firearms are probably the most obvious),... it

would be possible to predict that many of the Chinese colonists [in Inner Mongolia] would eventually convert themselves, as a result of the failure of their agriculture, into a pastoral people, and would finally, in order to give their pastures the necessary seasonal rotation in use, adopt a migration-cycle, thus becoming nomads and assimilating their society to that of the older nomad.[15]

It is no secret that nowadays Chinese state policy is to eradicate the pastoral nomadic way of life in areas ranging from Inner Mongolia to Tibet. In January 2008, for instance, it was reported that some 52,000 Tibetan herding households would be forcibly relocated and resettled into permanent housing.[16] This has been an unrelenting trend for decades, and has only accelerated in the post-Mao era of privatization.[17] Yet, it is also worth noting the emergence of a small, but discernible, voice from the scientific community in China (particularly Inner Mongolia) calling for a reversal of long-standing Chinese grassland policies.

While Owen Lattimore's wistful notion of Chinese colonists converting into pastoral nomads is most unlikely, three Chinese and Inner Mongolian authors of a 2007 article in the journal *Ecological Economics* did issue a call for the reversal of long-standing Chinese grassland policies in order to "correct irrational land use."[18] The authors of this article put the blame for Inner Mongolia's steppe-land degradation squarely on agricultural practices: "Agricultural output is not only low and variable, but farming methods in current use accelerate wind erosion and desertification."[19] Further expanding, the authors, two of whom have backgrounds in the environmental sciences, offered a stinging critique of Chinese policies:

It must be accepted that expansion of cropping into this fragile eco-system was an unfortunate mistake and that the long term sustainable management of these grassland eco-systems could benefit from a reversal of policies that are exacerbating the problems of land degradation and the adoption of land use practices that have been successfully applied for centuries by Mongolian herders.[20]

Hand-in-hand with a return to traditional pastoral nomadic practices and the banning of crop cultivation in Inner Mongolia, the authors call for the transfer of "[t]he surplus human population" out of Inner

Mongolia to Southeastern China. Those to be transferred (obviously Chinese farmers) are referred to as "ecological migrants."[21]

While the authors of this remarkable piece blame the grassland household contracting system that has been in effect since the 1980s as the prime factor in destroying traditional pastoral nomadic practices, it is clear from Owen Lattimore's writings that, to use Lattimore's own words, "there has for centuries been no true nomadism in Mongolia."[22] Lattimore wrote those words in 1935 in reference to the long-term, damaging effects of Manchu administrative constraints and the policies of the Chinese Republic on the Mongols of Inner Mongolia. In other words, a return to an idealized tradition of pastoral nomadism was, even as early as 1935, deemed an impossibility. Lattimore did, however, strongly believe in the flexibility of pastoral nomadism, a belief shared by many other experts in the field such as Anatoly Khazanov in his writings on comparative pastoral nomadic societies and Emanuel Marx in his work on the Bedouins in the Mideast.

Agriculture in the Socialist Era

If we examine Owen Lattimore's writings about agriculture in Mongolia some 30 years after his "The Geographical Factor in Mongol History" and "On the Wickedness of Being Nomads," it is clear that his 1961 visit to Mongolia inspired a revisiting of the issue of agriculture in Mongolian history. Lattimore urged historians to turn their research to the issue: "The alternation of periods of flourishing agriculture and abandoned agriculture is an important but hitherto neglected theme of Mongol history."[23] Reviewing the geographic zones of Mongolia where millet, wheat, oats, and even irrigated rice had been cultivated in past centuries, Lattimore wrote optimistically about the future of agriculture in Mongolia, particularly because in his view, writing in 1962, agriculture could coexist with the pastoral nomadic economy: "[a]griculture feeds livestock instead of displacing it."[24] By this, he meant that winter feed for livestock was a very significant segment of agricultural production.

Lattimore's appraisal that "the Mongols will be able to make farming a more important adjunct to livestock-raising"[25] echoed the goals of Mongolia's five-year plans in the socialist era of centralized economic planning. In 1972, for instance, the Mongolian People's

Republic (MPR) minister of agriculture claimed that 500–600 thousand tons of hay were produced annually in Mongolia and that winter fodder reserves had increased by one-and-a-half fold between 1962 and 1972.[26] As early as 1959, the Central Committee of the MPRP set the goal of "fully meeting the domestic demand for grain from an intensive development of grain farming"; this would be made possible by the development of so-called virgin lands.[27] In Selenge Aimag, for instance, the Seventh Five-Year Plan (1981–1985) called for the development of "about 60 thousand hectares of virgin land" along with a 12.7 percent increase in livestock herds.[28] Overall, state planning clearly was predicated on a transition away from nomadic agriculture and toward a fully mechanized agricultural sector.

Lattimore's optimism about the future of crop farming in Mongolia was not fully supported by another Western observer who had the opportunity to travel in the Mongolian countryside in the 1960s. Reginald Hibbert, who was the British resident Chargé d'Affaires in the MPR from 1964 to 1967, viewed the Soviet aid that led to the establishment of machine hay-cutting stations in the late 1930s as key to the future well-being of the livestock sector of the Mongolian economy: "This seems to have been the first attempt to tackle one of the fundamental and chronic weaknesses of Mongolian animal husbandry, the lack of winter fodder."[29] Hibbert, who observed several expansive wheat-growing regions in Bulgan Aimag in 1965, was positively impressed by the "Animal Husbandry Machinery Stations" that were heavily subsidized by Soviet aid. Hibbert's overall assessment of the MPR economy, however, was rather bleak: he saw the Soviet emphasis on agricultural production over animal husbandry as a product of the Soviet need for grain. Hibbert claimed that starting in the early 1960s, the MPR exported wheat to offset Soviet grain shortages. This is borne out by other sources.[30] Interestingly, Hibbert also witnessed thousands of herd animals (sheep, cows, yak, and horses) being "exported on the hoof" to the Soviet Union while he was touring northern Khövsgöl Aimag in 1965.[31] He further depicted the Soviets as diverting resources from "what is supposed to be the key sector of the economy, animal husbandry" to industry and grain farming.[32] In sum, Hibbert, in contrast to Owen Lattimore, believed that under the MPR development of the agricultural sector, while helpful to the livestock sector, was primarily focused upon fulfilling Soviet priorities and needs.

The question of agricultural productivity in socialist-era Mongolia, however, remains a complex question that economists, historians, and former Mongolian MPRP officials continue to debate. At the end of the socialist era, in 1989, crop production represented 30 percent of Mongolian "agricultural" production, the other 70 percent consisting of livestock production. Grains (mainly wheat) constituted 72 percent of crop production in 1989, with potatoes second at 19 percent.[33] A recent study of the 48 state farms that produced most of the grain and potatoes in the period from 1976 to 1989 concludes that, in spite of the substantial Soviet subsidies to the state farms, Mongolian grain and potato farming was marked by inefficiency and low productivity prior to the *perestroika*-inspired reforms of the 1980s. In Mongolia's last five-year plan (cut short in 1989), new incentive systems and more autonomy in management were introduced into the state farms, leading to much higher productivity on the eve of the transition to the free-market era.[34]

Even a former director of state farms, Tsogt-Ochiriin Lookhuuz (1923-), has provided a mixed account, often quite self-serving, of state farm finances in the socialist period.[35] While describing the state farms of the immediate post–World War II era negatively in terms of their inability to turn a profit, Lookhuuz credited himself with much success in later years in tying wages to productivity. Lookhuuz acknowledged, however, the importance of state subsidies in state farm operations. Even while touting the success of the state farms, Lookhuuz referred to rich pasturelands ruined in the 1970s and 1980s by conversion to agriculture in areas that were ill suited to crops.

With the total elimination of Soviet subsidies after 1989, with the dividing up of the former state farms, and with the "shock therapy" that Mongolia's economy as a whole was subjected to in the 1990s, agricultural production was hard hit in the first two decades of the postsocialist period. Is it possible that the socialist era (post–World War II) represented a high point in Mongolia's grain production in the cycle of what Lattimore long ago described as "alternating periods" of agricultural prosperity and downsizing or virtual disappearance of agriculture? Of course, cycles of agriculture in post-nomadic empire Mongolia would obviously derive from a variety of differing economic and political priorities, depending upon the land use policies of a particular ruling elite or governing body. It is nonetheless worth pointing out that by early 2008, Mongolia found itself in the position of importing most of its wheat and wheat-based products (breads, cereals) from Russia.[36] In fact, in April 2008, it was announced that Russia would export 100,000 tons of wheat to Mongolia (with export taxes being waived).[37] With a potential 35 percent increase in the cost of wheat-based products expected in 2008, the Mongolian government's focus on promoting domestic food production and reducing reliance on foreign imports was understandable.[38]

Periodic reports in the Mongolian press over the past 20 years vis-à-vis allegedly unhygienic Chinese food imports have led to a distinct preference for foods produced within Mongolia itself. In terms of vegetable production, Mongolia succeeded in producing 50 percent of the vegetables consumed in-country in 2009, though 100 percent of rice, vegetable oil, and fruits were imported, as well as more than 50 percent of wheat.[39] By the summer of 2010, optimistic reports of progress in liberating the Mongolian consumer from the necessity of buying Russian wheat and Chinese produce were circulating. Yet, even before the calamitous wildfires that hit Russia, which remained Mongolia's main source of wheat and flour imports, during the summer of 2010, Mongolia reportedly had already imported 64,000 tons of wheat and 70,000 tons of flour in 2010. Russian flour is considered higher quality than Mongolian flour by Mongolian consumers.[40] In the fall of 2010, however, with the agricultural harvest complete, the Ministry of Food, Agriculture, and Light Industry reported that Mongolia had become 100 percent self-sufficient in grains.[41] Credit was given to the government's campaign for self-sufficiency in foodstuffs. In spite of some success in achieving the goal of self-sufficiency in many food staples, Mongolia nonetheless remains quite dependent upon climatic and economic events beyond its borders and beyond its control.

Recent Developments in Agriculture

When we examine agricultural production in contemporary Mongolia, it is important to differentiate grain production for human consumption from the hay and fodder crops intended for livestock in the winter months. The state farm system in the socialist era developed both of these sectors intensively; it is estimated that the Mongolian state oversaw the production and distribution of over 200,000 tons of barley and oats, about 300,000 tons of so-called green fodder (annual and perennial plants), and 350,000 tons of sunflower and maize silage.[42] After 1990, Mongolia's herders had little choice but to return to traditional grazing strategies that have been described as "extensive" (as opposed to the "intensive" strategy of the state farms). From 1990 until very recently, there has apparently been very little cultivation of fodder crops.[43] There is not much data available concerning the extent of haymaking nowadays, but recent reports indicate that haymaking and hay storage are considered useful by herders only when geographically close to where the hay will be needed in winter and spring. Most hay production is geared toward consumption, not toward filling fodder reserves for emergency use.[44] While the number of winter/spring shelters for animals seems to be increasing, most livestock still must rely on grazing grasses and shrubs in pastures that have been "reserved" for the crucial winter and spring months.[45] The recent *zud* of winter/spring 2010 pressed home the need for herders to put up more forage for their wintering livestock in order to prepare for the worst.

In terms of agriculture for human consumption, opinion in Mongolia has certainly swung decisively toward the desirability of self-sufficiency in wheat production, vegetable production, and other food staples. Obviously, rice production is not feasible in Mongolia's climate, so at least one major component of the Mongolian diet will of necessity be imported, though not entirely from neighboring China. In the fall of 2010, Mongolia signed an agreement with Laos to lease 10,000 hectares (almost 25,000 acres) of land in Laos for rice production.[46] This venture is part of Mongolia's attempt to provide a reassuring semblance of self-sufficiency, even if the product is, in fact, imported from abroad. If one adheres to Lattimore's theory of a cyclical history of intermittent agriculture, then the postsocialist era thus far has proven to be an era of diminished agriculture; however, the new interest in self-sufficiency may result in a blossoming of the agricultural sector, climate change permitting.

How long the present phase of a return to traditional grazing practices will continue is open to speculation. It never ceases to amaze this observer that for a good part of the year (an estimated 180–200 days) much of the livestock in Mongolia relies on dried grasses on winter/spring pastures.[47] Mechanized haymaking is expensive. Foreign-funded rural development projects like the International Fund for Agricultural Development's (IFAD) ongoing Rural Poverty Reduction Programme have attempted to jump-start haymaking and vegetable cropping by providing small tractors, reapers, rakes, and other equipment to target areas (in this case, four aimags: Arkhangai, Bulgan, Khentii, and Khövsgöl). Mechanized hay production, the fencing of hayfields, and the training of horses, oxen, and camels as harness animals led to higher yields of hay in those targeted regions, as depicted in the 2005 Progress Report (the most recent available online).[48] Yet, extending the resulting benefits from a well-funded, geographically limited pilot program to the country at large is another issue (one of scale and resources). Periodically, at least some of Mongolia's herders are still forced to turn to the State Reserve Fund for emergency delivery of fodder even in non-zud years when severe snowstorms strike in the winter months.[49] In theory at least, all 21 aimags and some *sums* have hay and fodder reserves that are coordinated under the National Emergency Management Agency (NEMA).[50]

What are the future prospects for agriculture in Mongolia? In order to eliminate the enormous reliance on imported wheat and flour, the Mongolian government in February 2008 proclaimed the so-called Third Campaign for Reclaiming Virgin Lands (also referred to as ATAR-III), the first two such campaigns having occurred in 1959 and 1976.[51] As explained by then food and agriculture minister Ts. Gankhuyag, only previously plowed lands were to be reclaimed; no new lands were slated to be converted to crop use. Modeled on the socialist-era top-down model of economic planning, financing, and implementation, this campaign would include restoring agricultural machinery, importing large- and medium-sized tractors, and increasing irrigated lands. So-called micro agro parks were to be set up throughout the country, and these would entail government-funded greenhouses, storehouses, and seed supplies, along with technical assistance for citizens interested in engaging in agricultural production.[52]

The so-called Third Campaign, which ran from 2008 through 2010, aimed at total self-sufficiency in wheat and vegetable (including potato) production. Government assistance to farmers took the form of granting hundreds of tractors, seed-planting and potato-growing combines, and other agricultural machinery as well as tons of wheat and vegetable seeds and mineral fertilizers.[53]

Contradictory reports in the Mongolian press made it difficult to assess the results of this new campaign at the end of its first year. The January 2009 "Economic and Social Situation of Mongolia" (a periodically issued overview) seemed to credit the newly instituted campaign with increasing agricultural harvests in 2008.[54] Yet, the December 2008 meeting of the National Council for the Agricultural Development Program, Third Campaign for Reclaiming Virgin Land, chaired by Prime Minister S. Bayar, found that there had been no increase in vegetable production in 2008. Prime Minister Bayar reemphasized the goal of self-sufficiency in food supplies in Mongolia's aimags. Aimag officials were instructed to take a variety of actions including "promoting the development of greenhouse agricultural farming in settlement areas, increasing storage facilities for vegetables, renovating techniques and technology, and increasing VAT and Customs duty on imported vegetables."[55]

In October 2010, the Minister of Food, Agriculture, and Light Industry reported astoundingly good wheat harvests spanning 2008–2010. There was reportedly sufficient wheat for Mongolia's entire population.[56] While 100 percent of the demand for potatoes could be met, only 53.7 percent of the demand for vegetables would be met by the 2010 harvest.[57] The 2008–2010 ATAR-III Campaign is also credited with increasing ploughed lands, with a gain between 2009 and 2010 of 7,400 hectares (about 18,285 acres) devoted to agriculture. The ATAR-III Campaign also oversaw the modernization of an estimated 60–80 percent of agricultural techniques under government tutelage. Government support is crucial to the success of this sector, as the September 2011 trip of Prime Minister S. Batbold to Selenge and Töv Aimags highlighted. The Prime Minister presented ten combine harvesters imported from the United States to four aimags (Khövsgöl, Selenge, Bulgan, and Dornod) as well as assorted agricultural equipment to smaller enterprises engaged in vegetable and potato cultivation. The record grain harvest expected in fall 2011 was also credited to intensified government support of the agricultural sector.[58] With the conclusion of the Third Campaign for Reclaiming Virgin Lands, a new national program dubbed "Guarantee of Food Safety" was in the planning stages in late 2010.[59]

While such campaigns to promote agriculture have worthy goals, one also has to realize that in today's Mongolia, agriculture is only one piece in the larger land use jigsaw puzzle. In many cases, conflicting claims to land often remain unresolved because of ambiguities in the legal sphere. In addition to agriculture and herding, two other sectors—mining and, to a lesser extent, tourism—have made inroads upon the Mongolian landscape. The growth in acreage of protected areas has also had an impact: as of 2008, over 24 percent of Mongolia's land was under protected status through designation as Strictly Protected Areas, National Parks, Nature Reserves, Monuments, and aimag and sum protected areas.[60]

THE FUTURE OF SEMI-INTENSIVE DAIRY FARMING IN MONGOLIA

In Mongolia's peri-urban areas, in other words, the outlying regions that abut Ulaanbaatar, Darkhan, and Erdenet, competition for grazing lands has been intensifying in recent years. The competition is between semi-intensive dairy farms and traditional pastoral nomadic herders who have increasingly migrated closer to urban areas in search of education for their children and better access to markets for their livestock products.

Semi-intensive dairy farming has a history going back to the socialist era in Mongolia. In 1968, the first state dairy farm was established in Darkhan, and by the end of the socialist era in 1990, there were 43 state dairy farms that supplied 25 percent of all raw milk in Mongolia.[61] By 2008, there were some 900 dairy farms, mostly family owned, that were delivering raw milk to Ulaanbaatar. All these dairy farms were located in Töv and Selenge Aimags. While the Mongolian government's support for semi-intensive dairy farming has been hampered by fiscal constraints, dairy farms nonetheless have benefited from the resale through the Ministry of Food, Agriculture, and Light

Industry's Animal Husbandry Division of imported tractors to dairy farms at a reduced price.

With the population of Ulaanbaatar projected at 1,270,000 by 2015, the consumer demand for milk products will only increase. Unfortunately, the dairy-processing industry is underdeveloped and underused, in part because of uneven seasonal supplies of raw milk, the continuing Mongolian consumer preference for unprocessed milk and dairy products, and the pricing of processed milk products. In terms of mechanization, very few of today's dairy farms in Mongolia have milking machines; most farmers milk by hand. Farmers need silage harvesters for harvesting feed crops like corn and alfalfa. Dairy farmers either make their own hay or buy hay from Töv or Selenge Aimags.

From mid-May to mid-October, dairy cows are grazed in pastures in peri-urban areas where overgrazing is a major problem. The remainder of the year, however, dairy farm cows are housed in cow barns. The cow barns in Batsumber Sum in Töv Aimag that I visited in August 2011 were all modeled on 73-year-old Davaakhuu's pioneering family dairy farm, the first such of the postsocialist era. With design and construction aid from an Amish group and the University of Pennsylvania, Davaakhuu's barn was designed with a milk-cooling room: cold water drawn from a well keeps jugs of milk cold in preparation for collection by truck. Davaakhuu's farm is also the collection point to which other family dairy farms bring their milk. Davaakhuu himself gained expertise as a manager on a socialist state farm from the 1960s to 1980s, overseeing some 800 dairy cows. His farm now consists of ten Brown Swiss of Kazakh descent, downscaled from 24 head when he was younger. The barn itself and its many replicas in the region can accommodate up to 24 head (see figure 6.1).

The migration of herders into the same peri-urban areas utilized by semi-intensive dairy farmers has greatly exacerbated the environmental degradation of pastureland. While semi-intensive dairy farms have the right to fence pastureland, it is apparently quite rare for this to happen. Paragraph 52.5 of the 2002 Land Law states: "Pastureland fenced for purposes of developing intensified settled livestock breeding or farming of tamed animals can be given for use to citizens,

Figure 6.1 Davaakhuu at entrance to his model barn, Töv Aimag, 2011.

companies and organizations regardless of the season upon certain agreements and terms."[62] Recently, the Millennium Challenge Account (MCA) has initiated a project to support peri-urban dairy farms (the Peri-Urban Land Leasing Activity of the Property Rights Project). The MCA project will arrange long-term leases of about 300 tracts of pastureland near Ulaanbaatar, Darkhan, and Erdenet for selected "herder groups" in order to reduce pasture degradation.[63]

The overlapping spheres of sedentary-style dairy farming and traditional pastoral nomadic herding of livestock will continue to strain pasture and water sources in peri-urban areas. International development agencies in conjunction with the Mongolian government will play major roles in determining rights to land use for both types of livestock producers. In this author's opinion, both semi-intensive dairy farming and traditional livestock herding have much to contribute to Mongolia's food supply and economic development. If land use rights can successfully and fairly be negotiated, there is no reason why the two spheres cannot continue to coexist. Both spheres will also have to contend with ever-expanding mining operations and the development of tourist bases in the countryside.

AGRICULTURE, HERDING, MINING AND TOURISM: WHOSE LAND?

MINING

A prime example of a brewing conflict of interest over land use can be seen in the *ezemshigchiin gerchilgee* (certificates of possession) that have been issued in several if not all of Mongolia's aimags in recent years for winter/spring pastures versus mining permits. Sum governors were invested with the authority to issue certificates of possession to herders for individual campsites and for winter-spring pastures. Yet, aimag governors have the authority to issue mining licenses for exploration and exploitation, and the aimag authorities may be unaware of the prior issuing of possession certificates to herders in the same location.[64] The legal ambiguities that surround certificates of possession could be resolved in further amendments to the land law, and, ideally, the potentially contradictory lines of authority emanating from sum and aimag governors in their respective issuing of possession certificates and mining licenses need to be addressed.

Throughout mineral-rich Mongolia, conflict over land use is brewing. Even in northernmost Mongolia, where the Dukha reindeer herders graze their reindeer on the moss and lichen of the taiga, the discovery of gold in 2009–2010 in northern Khövsgöl Aimag led to physical confrontations resulting in several deaths, the misuse of reindeer for transportation to remote mines, and a clear disruption of traditional cultural patterns among the West Taiga Dukha.[65] East Taiga reindeer herders were headed west in summer 2010 to get rich quick in what promises to be an unraveling of the reindeer herders' unique way of life. In July 2010, in Uvs Aimag in Mongolia's northwest, security guards working for a Chinese gold mining company clashed with local residents who claimed that the company was polluting the Orlogo River, the area's main water source.[66] Some 300 herders and artisanal miners in Orlogo Bag demonstrated against the Chinese-owned mining company.[67]

Further examples of conflict between herders and mining operators spring up continually in the Mongolian press, as do accounts of pollution derived from mining. The effects of mining, particularly gold mining, are not geographically restricted, since gold and other minerals are plentiful throughout the country. In August 2010, for example, government inspectors found the Orkhon River basin in central Mongolia to be so heavily polluted by the numerous gold mining companies working in the region that all extraction work was ordered halted at least temporarily.[68] The Orkhon River is a culturally rich area of central Mongolia with plentiful "man stones" dating from the pre-Mongol Turkic and Uighur Empires; it is also an ecologically rich area supporting plentiful livestock herds that depend upon the waters of the Orkhon and its numerous tributaries.

In the southern part of Selenge Aimag, environmental activists associated with the Ongi River Movement shot at mining machinery and damaged vehicles belonging to a gold mining company that was illegally operating at the mouth of the Ongi River.[69] The Ongi River region has suffered enormous environmental damage as mining companies rely on river diversion and high-pressure water cannons to flatten hillsides. In the Ongi River basin area, some 60,000 nomadic households and one million head of livestock depend upon rivers and streams, many of which are running dry; the area's groundwater has suffered contamination from mercury and other chemicals—all results of gold mining operations.[70]

It is not just the large-scale mining companies, many with heavy foreign investment, that are to blame; so-called ninja or small-scale artisanal miners have also become the object of local herders' ire. These individual miners acquired the nickname of "ninja" miners because of the green, turtle shell–shaped mining pans that they wear on their backs. Examples of herder–ninja miner conflict abound. In Darvi Sum in Gov'-Altai Aimag in August 2010, 20 people suffered injuries and one man was killed as local herders clashed with some 100 ninja miners who were digging for gold in pastureland. The herders were reportedly outraged at the ninjas' destruction of the pastureland, and even though they were outnumbered by the ninjas, it was a herder who shot and killed a miner. The ninjas had refused to back off, claiming, "What else can we do? If we don't dig, we die of hunger."[71]

Numbering about 100,000 throughout Mongolia, ninja miners often move in to work sites abandoned or depleted by large-scale mining companies. Ninja miners are known to use mercury while panning, and thus their arrival in a locale is generally unwelcome. Many ninja miners had once been herders, perhaps forced out of herding by a zud, or possibly seeking a somewhat (though not much) easier way of life, or possibly lured by a get-rich-quick mentality that has characterized gold miners across continents and across centuries. The plight of artisanal miners has not gone unnoticed by foreign aid organizations. The Swiss Agency for Development and Cooperation (SDC), for instance, developed a mercury-free, small-scale ore-processing plant in Mandal Sum in Selenge Aimag in 2008.[72] In August 2010, the SDC transferred ownership of the plant to the local community company that operates under the aegis of the Sum Governor's Office. The artisanal miners employed at this plant will benefit from plant revenues directed toward the health and education of miner families. As well as legitimizing artisanal miners as an occupation group, this project also intends to clean up what has been an environmentally disastrous subindustry.

Overall, however, the scale of the environmental impact of mining is astounding: as a result of both company mining and ninja mining, some 850 rivers and 1,000 lakes throughout the country have dried up and chemical pollution of groundwater, lakes, and rivers is also on the rise.[73] The Oyu Tolgoi copper-gold project, located in Ömnögov' Aimag, contains an estimated 81 billion pounds of copper and 46 million ounces of gold. With the government of Mongolia holding a 34 percent stake in this, the world's largest undeveloped copper-gold mine, it is clear that Mongolia's economy will be increasingly dependent upon the mining sector in spite of its poor record of abiding by environmental regulations. In 2011, profits from coal mining exceeded profits from copper, and Mongolian coal exports replaced Australian coal exports as the largest supply source to China.[74] In future years, mining will undoubtedly contribute more to Mongolia's GDP than livestock herding (which is the main component in most listings of "agriculture"), and the Mongolian government has repeatedly identified the mining sector as the top priority in the country's economic development.

Mongolia's elected officials on the national and local level will face enormous temptations to favor mining over herding when land rights need to be adjudicated. As many observers have noted, enforcement of environmental regulations often fall to local sums that may have only one police officer with authority to make arrests. Since monetary fines for transgressors are minimal, and enforcement is spotty, controlling illegal mining activities remains a major challenge.[75] Government salaries are notoriously meager, and kickbacks, bribes, and other forms of corrupt exchange will undoubtedly tempt officials to favor mining rights over pastoral nomadic rights. Mongolia's herders have little surplus with which to "gift" their local representatives. Sharing the revenue from mining projects—as in the plan to distribute shares of, and annual dividends from, the Tavan Tolgoi coal mine in the South Gobi to every Mongolian citizen[76]—may buy some good will, but will that be at the expense of strict financial and environmental oversight and accountability? Large-scale mining and artisanal ninja mining will continue to compete with livestock herding for land and water for generations to come.

On the positive side, however, it is clear that Mongolia's government would not condone or resort to violence, as the government of China has, in resolving mining-herding disputes. In May 2011, hundreds of Inner Mongolian herders in China demonstrated at their local government offices to protest a Chinese coal trucker's killing of an Inner Mongolian herder and to protest against Chinese mining companies' displacement of herders and destruction of pastureland. Several protesters were reportedly beaten by Chinese police in Ujumchin, Inner Mongolia. In China, the government is the protector and patron of the mining industry. Inner Mongolia is China's top producer of coal and rare earth elements. Conversely, the government of China has shown no interest in preserving livestock herders' way of life; as is well known, protests of any sort in China are met with harsh crackdowns. To diffuse the anger of Inner Mongolian herders, the Chinese government summarily sentenced the offending coal truck driver who ran over the herder to the death sentence.[77] The people and government of Mongolia need only watch such events unfold in neighboring Inner Mongolia to see a negative model of conflict resolution among competing interests—mining and herding.

Tourism

Tourism in Mongolia is still in an evolutionary stage. A bit of historical background will explain how the industry has had to reshape its image in the postsocialist era. Tourism in socialist-era Mongolia was operated as a state monopoly. Mongolia's Juulchin travel organization was founded in 1954, and received its first group of visitors from the West in 1962. Juulchin was patterned on the Soviet Union's Intourist Agency, and, indeed, Juulchin guides were often sent to the USSR for training. The overlap between the intelligence sector and the hospitality sector was well known in both the USSR and the MPR. Not surprisingly, Juulchin became notorious for its often inept and occasionally rough handling of tourists from the "capitalist countries." Stern lectures on the superiority and achievements of socialism were common as the author can attest, having been "welcomed" by Juulchin in 1978, 1979, and 1983.

Tourism since 1990 in Mongolia has shed its monolithic structure, and a plethora of tour companies now operate in-country. The Mongolian government does, however, stay involved in tourism by drawing up policies to promote this important sector and to safeguard the environment. Before 2008, responsibility for tourism fell to the Ministry of Roads, Transport, and Tourism, but during 2008, tourism was attached to the newly constituted Ministry of Nature, Environment, and Tourism, thereby acknowledging the important link between tourism and the land.

Tourism presents an opportunity to engage in environmental conservation (so-called ecotourism) but may also present a hazard to the integrity of the landscape and to nomadic culture that constitute the main reasons why tourists visit Mongolia. Ecotourism has come in for its share of well-deserved criticism as "a trendy, catch-all word applied to almost any activity that links tourism and Nature" and as "a gimmicky marketing tool... that hides irresponsible, unethical and unsustainable practices" in worst case scenarios.[78] As if to echo this sentiment, the director of Mongolia's National Tourism Center, which is a government agency, N. Molor, lamented in August 2010, that "the number of tourists has only meant a rise in environmental pollution and exploitation of natural resources." He added that "local communities reap little benefit from tourism in their area, and are thus unable to work for regional development with the revenue that properly planned tourism should generate."[79] Yet, the director puzzled over the problem of involving more local people as guides when language is an issue: English, for instance, may be increasingly common as a second language in Ulaanbaatar, but it is oftentimes poorly taught if taught at all in rural schools. Molor also expressed frustration with the widespread use of the term "ecotourism" as a ploy to increase business among tour operators.

The development of tourism in Mongolia has been aided by the generally correct conception of Mongolia as a safe country for foreigners. While crime is a problem in Ulaanbaatar, the Mongolian countryside remains relatively safe. Yet, tourism's expansion will continue to be hindered by the extremely short travel season—May to September, the warmer months when the countryside is more accessible—although some attempt has been made to attract foreign tourists to camel festivals in the South Gobi in the depths of winter. Even with tourism's natural limits, tourist-derived income in 2006 constituted 10 percent of Mongolia's gross national product, and all indicators point to the numbers of foreign visitors continuing to rise as they have fairly consistently risen over the past two decades.[80] With the exception of 2003 when a SARS outbreak kept visitors away from most Asian countries, the years 1998–2006 saw foreign tourist visits to Mongolia grow at an average rate of 13 percent a year.[81] In spite of its steady growth, tourism faces challenges related to its haphazard, largely unregulated, and demand-driven development.

To understand the positive and negative impacts on herders and the land, we will examine a few examples of commercial tourist operations in Mongolia. It will become clear that tourism has great potential to leave its imprint upon the environment, culture, and economy of the countryside in a variety of ways. First, the Gün-Galuut Nature Reserve located south of Baganuur in Töv Aimag in central Mongolia, some 130 kilometers southeast of the capital, hosts an annual two-day "Nomads' Day Cultural Festival."[82] The Reserve itself, which was established in 2003, maintains a website that gives a detailed description of its mission: it "focuses on mitigating adverse impacts to biodiversity and natural integrity of the reserve and developing ecotourism." The Gün-Galuut Community Association offers a pricelist for activities for "domestic and international guests," suggesting that Mongolian urbanites from Ulaanbaatar might be among those who wish to ride a horse, camel, or yak, or perhaps go fishing with a local guide. Also offered is "staying at a nomadic family" for ten dollars, presumably mainly of interest to foreigners. This homestay-for-money reflects an interesting shift in traditional Mongolian concepts of hospitality, that is, opening one's *ger* to the occasional traveler in need of lodging. In the age of market relations, commercial transactions have of necessity supplanted the old norms of hospitality, particularly where foreign tourists with dollars are concerned. Of course, such transactions—paying for a local guide, paying for a ger homestay, paying for a horse ride—all benefit those local nomadic families who may thus supplement their livelihoods at least during the all-too-brief tourist season.

In 2004, the Steppe Nomads Tourist Camp was founded on the banks of the Kerülen River in the Gün-Galuut Nature Reserve as the only ger camp in the Reserve. Every year on September 17 and 18 the "Nomads' Day Cultural Festival" attracts foreign tourists who, according to the official website of Mongolia's Ministry of Nature, Environment, and Tourism, "meet nomads in traditional costumes and take part in traditional activities, witness nomadic children showing off their mental abilities, observe the contest of the most beautiful, traditionally-dressed nomadic groups and try mouth-watering steppe delicacies."[83] A mini-*Naadam* (Mongolian festival featuring the three "manly sports" of wrestling, archery, and horse racing) is staged with tourists participating. This annual festival has both private and governmental sponsorship: sponsors include the Selena Travel Group, which is a Mongolian tourist company, the Office of the Governor of Bayandelger Sum, the "Natives' Council of Bayandelger Sum", the Gün-Galuut Community Association, and the Steppe Nomads Tourist Camp.

Undoubtedly, a windfall for those local nomadic herder families who are involved in the festival, such an event may also pose a downside. Are Mongolia's nomads, at least those who herd their livestock along the more traveled routes that allow tourist access, in danger of becoming a "living museum" of exotic and nearly extinct species? Is the integrity of nomadic culture affected when foreign tourists pay for a homestay in a ger? Consider this scenario, recurrent in my travels through several aimags in Mongolia: the vehicle with a foreigner arrives unannounced at a ger; once the family notices the foreign guest, a hustle-bustle of clearing out unwashed pots and plates, evicting noisy children and perhaps other guests, and a frantic attempt to tidy up the interior—all within a minute or two. Luckily this is not always the scenario, but it is a routine often enough encountered that leads one to question how well a foreign visitor can access traditional nomadic culture without, in the process, reordering its own routines and identity (see figure 6.2).

A Mongolian researcher in development policy polled opinions among rural residents in Övörkhangai Aimag vis-à-vis tourism's effects on culture. Interestingly, her respondents overwhelmingly saw positive effects: since many traditional religious and cultural practices had been suppressed in the six decades of socialist rule, they saw an incentive to revive such practices to share with tourists. Tourism thus has spurred a revival—in undoubtedly much revised forms, given the long hiatus—in areas like traditional music and dance.[84]

Nomadic Expeditions is positioned at the high end of environmental and cultural conservation in the Mongolian tourist industry. In business since 1992 and founded and managed by a Mongolian-American who has consulted with the government of

Figure 6.2 Ger interior, Zavkhan Aimag, 2005: A rare instance of a family routine uninterrupted by visitors.

Mongolia on the development of tourism, Nomadic Expeditions sets a high bar for other tour companies to match.[85] (A necessary disclosure here: I have used Nomadic Expeditions guides on five trips over the past ten years.) Nomadic Expeditions' predeparture information, for instance, urges travelers to shop for gifts in tourist ger camp shops, not in Ulaanbaatar, since local herder families will directly benefit from purchases of their crafts.

Nomadic Expeditions puts into practice what its literature promotes in terms of caring for the environment and sustaining the pastoral nomadic way of life through its starship ger camp, the Three Camel Lodge, located in Ömnögov' Aimag. The Lodge's website mentions a "first-of-its-kind cooperative agreement with the Bulgan Sum Township and Gobi-Gurvansaikhan National Park" in preventing poaching of wildlife and illegal removal of dinosaur fossils; it also claims to be an "active supporter" of the Ongi River Movement that seeks to protect the Gobi's Ongi River from the mining industry's effects. Along with its "No Plastic Bags in the Gobi" campaign, the Lodge recycles its organic waste for local farms to use as fertilizer. Through its support of local chicken, egg, and produce farms, the Lodge has certainly reduced the amount of food trucked in from afar. The Lodge also has its own greenhouse on site. With bragging rights to being perhaps the only tourist ger camp that separates all its trash, recycles, and transports unrecyclable items to designated dumps nearby, the Three Camel Lodge has by implication highlighted a growing problem at tourist ger camps in the Mongolian countryside. It is not uncommon to see an open-pit garbage dump on the perimeter of a ger camp if one takes a stroll of a few minutes. Reportedly, at Khövsgöl Lake National Park in the north of Mongolia, the use of biodegradable bags to collect and carry out solid waste has been introduced, certainly a step in the right direction.[86]

In terms of power sources, the Three Camel Lodge produces its own electricity with solar power with a windmill behind the Lodge providing supplementary wind-powered electricity. It is increasingly common nowadays to see solar panels and windmills standing next to individual herder gers in the Gobi and elsewhere in Mongolia. Certainly, the Gobi is not lacking in wind or sun! Mongolia's tourist industry as it operates in the countryside has an opportunity to take the lead in using clean sources of energy, in recycling, and in using responsible means of garbage removal. One hopes that tourism will continue to develop along these lines.

How might tourism truly benefit local herders and sum residents? Taking the example of Nomadic Expeditions, its Three Camel Lodge is making the effort to hire its employees from among South Gobi residents: in fact, the Lodge claims to be one of the largest employers in the Gobi, although this will certainly change as the mining sector expands. The Lodge has also signed on to the "Pack for a Purpose" nonprofit organization that guides tourists in purchasing and packing needed items that they then bring in person to local projects overseas. As the only such participant in all of Mongolia, the Three Camel Lodge lists projects that would benefit from specific supplies that tourists could pack along, including the town of Bulgan's school and hospital.[87] Mongolia's tourist industry at large could do far more to serve in a helping capacity, as employers of local guides, purchasers of local goods and services, and intermediaries in bringing together foreign tourists who may wish to contribute to rural educational, medical, and environmental establishments that have specified areas of need.

Ecotourism may also benefit conservation as in the example of the Ikh Nart Nature Reserve that straddles two sums (Dalanjargal and Airag) in Dornogov' Aimag. Since 2008, the Nomadic Journeys Tour Company has operated an "eco-tourist ger camp" inside the Reserve. The tour company benefits from holding the exclusive rights for overseeing foreign tourism at the Reserve; in return, a conservation head tax is collected and designated in support of the Ikh Nart Nature Reserve.[88]

Overall, tourism has the negative potential of competing with the herder population over scarce resources like water. Tourist operations have damaged the landscape in some areas of the country through vehicle tires tearing up pastureland. Overuse of timber from Mongolia's ever-diminishing forests as well as the still largely unaddressed problem of properly disposing of solid wastes at tourist sites in the countryside are also negative consequences. Environmental damage through vehicle parking and waste disposal problems at the mineral hot springs in Hujirt in Övörkhangai Aimag has been cited as a particularly bad example.[89] Tourism's demands on the land are numerous, but with careful regulation and management, tourism may ultimately prove to be a good companion to the herders' way of life.

Impact of Foreign Aid in Land Use Issues

Mongolia sometimes seems to function as a lab for foreign aid projects. In projects ranging from the "Mongolian Potato Program" (SDC) to the Global Environmental Facility (GEF) Small Grants Programme under the United Nations Development Programme (UNDP), we see foreign-funded efforts to integrate local livelihoods with preservation of natural resources as well as efforts to increase Mongolia's capacity to grow its own crops for livestock and human consumption. While international donor organizations have been criticized for their "piecemeal programs meant to assist herders," with questions raised concerning long-term viability and cost, it is nonetheless clear that piecemeal is preferable to no meal.[90] We will spotlight tourism and agriculture as two important foci of international aid donors working in the Mongolian countryside.

UNDP/GEF projects that promote tourism are not, initially at least, as beholden to the market economy, though ultimately the seed money that permits community-based tourism development will have to be replaced by monies earned by attracting tourists and competing in a very competitive tourist industry. Perhaps the best example of a successful tourist base camp that was initiated by foreign enterprise but is now a largely Mongolian operation is the Khustai Nuruu Tourist Camp in Töv Aimag. This tourist ger camp, only about two hours by car from Ulaanbaatar, is a favorite destination for foreign tourists who enjoy viewing the *takhi* (indigenous Mongolian wild horses, sometimes referred to as Przewalski horses) that come down from the surrounding hillsides to drink out of valley streams late in the day.

The Dutch Foundation for the Preservation and Protection of the Przewalski Horse, with funding from the World Wildlife Fund, was able to buy from zoos around the world some of the nearly extinct takhi, the only truly wild horse in the world, and, beginning in 1992, reintroduce them into Mongolia where they had been extinct since the 1960s.[91] A 50,000-acre nature preserve (Khustai National Park as of 1997) was established for the reintroduced takhi; this entailed the gradual removal of nomads and their livestock and a prohibition on hunting. To provide recompensation and means of further livelihood to the nomads who lost access to the land of the new national park, the UNDP and the Dutch Ministry of International Cooperation funded new cheese- and yogurt- production factories in nearby towns and also repaired old water wells and constructed new wells outside the park so that the displaced nomads would be able to provide for their livestock outside of park boundaries.

The Khustai National Park center hosts a tourist ger camp that provides employment to local residents, but the higher level jobs of park rangers (many of whom speak English) and research biologists are filled by Mongols with more advanced education. This ger camp has been, however, cited as a "model" for local community-based tourism. The Khustai National Park's ecotourism model has been praised for "providing a greater return than found in most protected areas of Mongolia."[92]

Another foreign-funded venture, this one supported by the UNDP's GEF Small Grants Program, provided funds in 2006–2007 to train local herders in Gov'-Altai Aimag in developing small-scale tourism: a tourist ger camp; guided tours to natural attractions; a village souvenir shop; and camel and horse tours. The motivation behind this project was twofold: to provide a "valuable new income-generating alternative" for herders who had suffered in recent decades from the vagaries of climate change; and the need to structure tourist visits to a natural site, Eej Khairkhan Mountain Natural Monument, which is much visited by domestic and foreign tourists. This project, according to the vice governor of Gov'-Altai Aimag, created ten jobs, which may not sound that impressive on paper, but likely made an important contribution to the budgets of at least a few of the pastoral nomadic households in the surrounding area.[93]

Another UNDP project conducted in conjunction with the Mongolian Ministry of Environment from 2005 to the present seeks to involve and train herders in the Altai Sayan Ecoregion Project: herders in the Altai Mountains keep data on endangered plants and wildlife that aids both conservation efforts and efforts to determine appropriate hunting. Foreign hunting groups provide a substantial amount of tourist income in Mongolia.[94] Herders who participate are encouraged to register as "sole users of natural resources" in their locales; the participating herders have also been encouraged to venture into small-scale tourism—setting up ger camps, offering horse or camel rides, and selling their own handicrafts.

Foreign aid projects have been crucial to Mongolia's development in the agricultural sector. One may surmise that the donor nations, all with industrial and agricultural sectors, and the donor organizations, all with a bias toward cropland use over livestock herding, have favored agricultural cropping over pastoral livestock herding. Yet, many of the projects are meant as auxiliaries to herding, not replacements per se. For instance, the GEF Small Grants Program has initiated projects that encourage the cultivation of *tsargas* (a Mongolian fodder plant that thrives on degraded and abandoned pastureland) in order to prevent further soil erosion, as well as projects to support the cultivation of buckthorn and black currant fruit trees on abandoned lands.[95]

The cultivation of potatoes, referred to as Mongolia's "second bread," has been a major focus of the SDC in the Mongolian countryside. From 2004 to 2010, the SDC's "Revitalization of the Mongolian Potato Sector Program" concentrated on identifying and distributing high-quality potato seeds suitable for Mongolia, improving production technology, expanding collaborative efforts in rural areas, and increasing profits from growing potatoes.[96] With imported German washing, drying, and packaging equipment, affordable, packaged potatoes that have been grown in various aimags in Mongolia are now being sold in Ulaanbaatar's larger markets.

The Mercy Corps, a secular aid organization working in more than 40 countries worldwide, has focused on the livelihoods of herders in the Gobi aimags in a variety of projects including the development of a forage monitoring system that assesses current and forecasted livestock forage conditions to aid herders in planning.[97] The Mercy Corps' emphasis on diversifying Gobi herders' income sources away from a dependence upon cashmere and wool market prices has apparently met with success; in 2010, the Mercy Corps announced that it was expanding its Market Opportunities for Rural

Entrepreneurs (MORE) into three additional aimags (Bayan-Ölgii, Dornod, and Khövsgöl). By the end of 2011, the MORE project expected to work in a total of 15 aimags to increase herders' and ex-herders' income through expanding access to markets and developing small businesses.

The SDC's Green Gold Pasture Ecosystem Management Project has been working along several fronts to fight pasture deterioration from overgrazing; it has also helped prepare the draft of the new law on pastureland.[98] On the local level, this project has supported the formation of Pasture-User Groups (PUGs) in which herders at the sum level, in conjunction with local governments, enforce sum land-management plans and land use regulations. This may include creating reserve pastures and wells, lands fenced for irrigated hay crops, and mutually agreed upon pasture rotation. The central idea behind the PUGs is to reduce conflict over land rights and to reduce overgrazing, problems that stem from unregulated open access to pastureland. Reserving pastureland for winter and spring use has been key to this effort. While the SDC reports the successes of the PUGs on the local level, there remains, according to the SDC's 2009 Annual Report, no "clear legal framework and strong political will to regulate grazing."[99] The PUGs have limited legal authority since national legislation on pasture management organizations has not yet been formulated. The SDC deemed the initial Green Gold pilot program (2004–2008) in which 400 PUGs with some 10,000 herder households participated so successful that an additional 40 sums in western Mongolia were to be added to the project.

Western Missionary organizations have also funded poverty alleviation and development programs both in Ulaanbaatar and in the Mongolian countryside. While evangelizing remains the primary goal of Christian organizations working in Mongolia, their humanitarian work in the urban and rural sectors has been significant. The Norwegian Lutheran Mission, for instance, ran a Selenge Development Project from 1995 to 2008, a project that included under its broad umbrella working with herders to breed fine- and semi-fine-wool-producing sheep, wool-processing, and family gardening centers. Most Christian evangelical organizations are rooted in Ulaanbaatar where they find plenty to work on: the demands for poverty alleviation have grown commensurate with the city's exploding population.

While one may indeed question the ultimate efficacy of numerous aid projects funded by a variety of organizations and nations, none of the projects that focus on the Mongolian rural sector seek to remove herders from their occupations if they desire to continue their traditional way of life. Rather, several of these projects aim to improve livestock breeds, improve market access, develop forage crops as protection against winter conditions, and offer supplementary incomes through crop agriculture and tourism. There were more tractors on the landscape of rural Mongolia in the summers of 2010 and 2011 than I had ever seen in previous visits. Mechanization has returned to the Mongolian countryside, thanks in part to the government of Mongolia's emphasis on agricultural self-sufficiency, and thanks in part to foreign aid projects. Further legislative efforts to recognize and coordinate land-management plans that have already sprung up across the Mongolian countryside would give herders more confidence in their rights to control access to their seasonal pastures and water sources.

As pastoral regions throughout Inner Asia come under increasing strain from development (mining and other forms of resource extraction) as well as from climate change, especially desertification, Owen Lattimore's observations on the complex interrelationship between pastoral nomadism and agricultural production ring with relevance. The land in Mongolia is under intense and complicated pressures from competitive and oftentimes noncompromising sources. One hopes that a deeper understanding of the past history of land use will provide some guideposts for the future.

CHAPTER 7

CONCLUDING THOUGHTS

HERDING IS A RISKY BUSINESS

Most observers can readily fathom the risks of livestock herding in Mongolia. In years when *zud* strike, the losses in animals and the resulting hardships for herder families are immense. The tremendous losses between 1999 and 2002 (over 11 million head of livestock lost in successive zud) and in the winter of 2010 (over 12 million head of livestock lost)[1] have resulted in ongoing debates over the viability and future of pastoral nomadic herding in Mongolia. In chapter 5, we took a brief look at a livestock insurance project (IBLIP or Index-Based Livestock Insurance Project of the World Bank and the government of Mongolia) aimed at reducing risk. This may be the most ambitious and potentially most effective approach to dealing with variables such as climate change, pastureland quality, and the market for animal products, all of which are part and parcel of pastoral nomadic herding.

In spite of the anticipated difficulties in enlisting herder participation in this project, particularly when animal product prices were on the decline and the cost of living on the increase, over the course of five years (2005–2009 inclusive) more than 15,000 herders insured more than 2 million head of livestock.[2] Other challenges include the high cost of the necessary midyear animal census conducted by the National Statistical Office (NSO), the need for committed long-term insurers after the pilot project ends, and the need to develop legal protections for insurers and insured alike. Yet, in 2009 alone, there was a 30 percent increase over 2008 livestock insurance sales. In 2009,

2,117 herders received payments totaling US$275,700 for livestock losses. Because the winter months of 2010–2011 were milder with fewer livestock losses, Mongolian insurance companies were able to reap profits.[3]

At the very start of the IBLIP Project, one historian dismissed "insurance schemes (which many herders either cannot afford or cannot understand)."[4] It is easy to underestimate the practicality and diligence of most Mongolian livestock herders. If herders could hold on to their way of life under socialism, they assuredly can embrace "insurance schemes" that are designed to bolster and perpetuate their livelihoods.

Analogous insurance projects have been implemented in other developing countries. For instance, in the Baltistan region of Pakistan, a communally funded insurance program has been instituted to insure herders' flocks of goats and sheep against losses due to snow leopard attacks. This project's mission is not only to secure the livelihoods of the local agropastoralists through the creation of a collective insurance fund to replace animal losses, but also to reduce the level of hostility toward this endangered species through a second fund based on ecotourism activities involving the snow leopard.[5] The founder of Project Snow Leopard, Shafqat Hussain, funds roughly half of the insurance project through annual premiums paid by herders; the other half is funded through his own ecotourism trekking outfit in the region.

While this example is geographically limited to a rather confined region that encompasses snow leopard habitat in northern Pakistan and involves relatively few herders as compared to the landmass of Mongolia and its far more numerous herding population, the Baltistan project raises an intriguing idea. Could the increasingly profitable tourist industry in Mongolia be brought into the livestock insurance equation? It is safe to generalize that tourists go to Mongolia not to see the sprawling, polluted city of Ulaanbaatar, but rather to see the countryside, particularly the pastoral nomadic way of life. Even in remote areas of Mongolia, herding households find their daily routines interrupted by foreign visitors arriving in jeeps or vans and hoping to be treated to the legendary nomadic hospitality. Would it be implausible to levy a head tax on international tourists who spend time among Mongolia's herders, and deposit this sum into the Livestock Insurance

Indemnity Pool (LIIP), thereby augmenting the fund that pays out for herders' losses?

As we have seen in chapter 6, a conservation head tax is already levied on international tourists who visit the Ikh Nart Nature Reserve in Dornogov' Aimag. While some funds should continue to be earmarked for conservation (which is, of course, an equally worthy cause), this alternative scheme would address herders' specific needs for a more secure livelihood. Tourism is now intertwined with the pastoral nomadic way of life in Mongolia, as we have seen in the examples provided in chapter 6. Linking the two economically is a concept worth considering.

To those who might object that pastoral nomads are in danger of becoming "museum pieces," one can only respond that in all likelihood the future of pastoral nomadism will require economic success based not only upon the economics of producing and marketing meat, dairy, cashmere, camel's wool, and other animal products, but also upon the economics of marketing a way of life that has always had enormous appeal to the outside world. The renowned anthropologist and researcher of nomadic peoples, Anatoly M. Khazanov, has long called for pastoral nomadic groups to modernize by producing for the commercial market and to turn to secondary pursuits to augment their economies. Yet, Khazanov specifically warned against tourist activities when he wrote that "pastoralists face the risk of being further marginalized and encapsulated, or of becoming zoo groups, the exotics for urban romantics and tourists for whom the sight of black tents represents a living museum's inventory."[6] In Khazanov's view, tourism waters down the integrity of the nomadic way of life. But, over the past 150 or so years of Mongolian history, a vast array of travelers, researchers, and tourists have threaded their way across the deserts, steppes, and mountains of Mongolia to observe, study, and write about the unusual way of life that they were fortunate enough to experience. Sometimes a thin line separates a tourist from a researcher as the diaries and journals of travelers a century or so ago suggest; would not those travelers of yesteryear be described as tourists nowadays?

From months-long visits to merely a few days, travelers and tourists have indeed enjoyed (and at times overextended) the hospitality of countryside Mongols. But the Mongols themselves have attempted to shape the impressions that foreigners might form. Showcase *ger* encampments and model state farms were already a part of the Sovietized experience in the 1960s, as the letters of Reginald Hibbert, the British Chargé d'Affaires at Ulaanbaatar, suggest.[7] On my first travel in the Mongolian countryside in 1978, it was crystal clear that preselected herder households were on the itinerary. A model herder ger surrounded by herds of horses (near Hujirt, central Mongolia) or camels (in the South Gobi) demonstrated all the achievements that the socialist system trumpeted. In 2010, I experienced an arranged visit to a nomadic family near Khongoryn Els in the Gobi that was eerily reminiscent of my socialist-era visits (1978, 1979, 1983, and 1987). The nomadic family was a "company herder" household under contract with the tourist company to provide camel *ayiragh* (fermented milk) and camel rides to foreign visitors. This was a direct descendant of the socialist-era practice, but also different in important ways. The 2010 "company herder" family spoke openly without any fear of censorship of the difficulties of winter 2010, their losses of livestock, and the need for a parallel income from hosting tourists. At the end of the visit, the wife and daughter of the herder offered handmade trinkets for sale: bracelets, tiny camel statues made of local sandalwood, wool slippers, and other items.

Figure 7.1 Motel and café (*guanz*) on the road in Arkhangai Aimag, 2002.

Nowadays, Khazanov's "urban romantics" from the United States, Japan, and other urbanized parts of the world do indeed crave the blue skies and stretched horizons of the Mongolian countryside. As the numbers of foreign tourists increase, the opportunities for commercial profits have grown apace. In an overland trip through Arkhangai Aimag along the so-called Millennium Highway in summer 2002, I was greatly surprised at the number of "for-pay" refreshment establishments (see figure 7.1).

Guanz (restaurants) in the form of small wooden shacks offering fast food made stopping in at a herder's ger obsolete in this part of Arkhangai Aimag. The Mongolian word "guanz" is a direct loan word from the Chinese word for "restaurant" or *guanzi*; thus, conceptually as well as linguistically, the Mongols have borrowed from the Chinese as the trend toward commercialization of the countryside accelerates. This type of small-scale commercialization, which is often a real eyesore on the landscape, will not necessarily benefit herding households, however. One would have to determine who runs the guanz and whether the profits are returned to herding communities.

Figure 7.2 Roadside vendors along a popular tourist route, South Gobi, 2010.

Rather than subvert traditional modes of hospitality by encouraging herders to start charging the drop-in visitor for a bowl of *süütei tsai* (milky salty tea), it would seem preferable to institute an indirect system that would allow herders to profit from tourism. A few individual tour companies, as noted earlier, already keep some herder families on contract. In addition, tour companies or individual tour guides could pay a surtax or fee that would increase the insurance fund that indemnifies herders for losses in the periodic zud that strike. Ultimately, however, it will be up to the herders themselves to decide how best to adapt to the market forces that serve both to include them in the globalization process and to raise questions about the sustainability of traditional cultural modes of behavior.

Democracy, Foreigners, and the Land

Open and transparent commercial and political transactions are vital to the further development of Mongolia's democratic system. Land has been subject to commercial fraud in Ulaanbaatar with the illegal sale of land in city parks, river floodplains, and protected areas. Reportedly, such illegal sales often result in the construction of houses and tidy profits for the businessmen and politicians involved. Ulaanbaatar land fraud was cited in a recent editorial in the Mongolian press as "just one example of how those who come to power... [are] making our democracy valueless and turning Mongolian laws into nothing."[8] In a sardonic conclusion, the editorialist speculates that it would be cheaper just to have one dictator instead of so many dictators who are responsible to no constituency. Yet, the final plea of this Mongolian writer is straightforward: "There is an urgent historical need for Mongolia to guarantee its democratic accomplishments by economic development. We want to live in [a] democratic and developed country."[9]

How will Mongolia's democratic system address the issues of corruption and profiteering? It is not yet clear that democratic values have translated into enforceable regulations and statutes to curb profiteering. As mining interests increase their reach, land in the countryside may become equally susceptible to gross manipulation by a small minority seeking immense profits unless safeguards are imposed.

Land fraud is not just an urban problem, though it is better documented in Ulaanbaatar than elsewhere.

The economic boom on the horizon includes a projected quadrupling of Gross Domestic Product (GDP) per head by 2018 (from US$2,000 in 2010). The anticipated economic surge is directly linked to two South Gobi mines: Oyu Tolgoi (copper, gold, silver) and Tavan Tolgoi (coal).[10] How will this infusion of wealth alter the lives of Mongolia's herders? Mongolia's president, Tsakhiagiin Elbegdorj, expressed great concern over the high risk posed by such a rapid economic shift: "If we get much more income and much more profit in a bad system with bad governance, I think Mongolia is in trouble."[11] Elbegdorj forecast mining's current 70 percent contribution to Mongolia's output as dropping to about 20 percent within the next 20 years. While a more diversified economy would prevent overreliance on world prices for just a few commodities (gold, copper, coal), one may also envision a drastic environmental impact during the projected two decades of mining's supremacy. Even with anticorruption legislation being enacted, the rural landscape and its available resources will be altered, divvied up, reapportioned, and perhaps left in ruins.

The Mongolian government is committed to distributing the wealth derived from mining. Mongolia's citizens—including herders—in 2011 were already receiving monthly cash payments from a national fund, the Human Development Fund, constituted from mining royalties and tax revenues. In late 2011, the monthly allotment to each citizen was 21,000 *tögrögs*, while Ömnögov' Aimag residents received an additional 500,000 tögrögs as a "local grant." One observer noted that the "local grant" seemed to make Ömnögov' residents less interested in taking construction jobs related to the mining industry and its infrastructure development in the aimag; instead, workers from other aimags, and in particular, skilled Chinese laborers were moving in to fill the new jobs.[12] Not only will cash handouts not solve Mongolia's problems with unemployment and lack of skilled labor, but cash handouts will also certainly not address problems of pastureland and water contamination or access, problems directly related to mining. The traditional pastoral nomadic way of life will still face enormous challenges.

A recent article, however, makes the case that at least some of the wealth derived from the mining sector could be earmarked to revitalize the herding sector.[13] Such a scenario would, of course, depend upon governmental activism and dedication to the long-term survival of pastoral nomadic herding. Government monies could be directed to increased fodder storage, well digging, pastureland management, and so on. In other words, mining does not have to pose a direct threat to herding on all fronts. If Mongolia's elected officials have a clear mandate to use reallocated funds derived from the mining sector to benefit the development of pastoral nomadic herding, one could envision a realignment of interests across the board.

Given the constant and growing demand for Mongolia's mineral resources—particularly the demand for coal by China—it would take a highly principled government to earmark revenues for the expansion of nonmining sectors, including herding. Herders, for the foreseeable future, may have to develop a louder voice and more expertise and clout in finance and policy making from the local to the national level.

Mongolia, a democracy, lives next door to a one-party ("Communist") state—China—that is immensely invested in Mongolia's economy and its land, primarily through mining rights. Anti-Chinese sentiment is nothing new in Mongolia, but a recent article by a French anthropologist has attempted to sort through the reasons behind the barely disguised anti-Chinese bias that is so common among Mongols today.[14] Examining rumors, ghost stories, and other forms of popular, oral narrative, the researcher points to the historical memory of "the harsh mercantile ethics of Chinese merchants" in earlier centuries (eighteenth to early twentieth) becoming conflated with the activities of Chinese in Mongolia today.[15] In other words, Chinese colonial domination of the Qing era, highlighted in both socialist-era schools and in postsocialist education, combined with the recent flood of Chinese businessmen into Mongolia, has resulted in a portrayal of the Chinese as parasitic—extracting vital resources from Mongolian soil to serve the needs of the 1.3 billion Chinese consumers.

How accurate is this perception? While Chinese businessmen, investors, and workers do not represent a Chinese state-directed solid economic phalanx, there is certainly enormous indirect state support for resource extraction from Mongolia. As a policy analyst recently wrote, "Beijing has identified Mongolia as a first destination for its 'Go Out' policy (Chinese: Zou Chuqu Zhanlue), by which Beijing hopes to secure access to natural resources while expanding China's political influence."[16] Not only is Mongolia geographically convenient from the Chinese state perspective, but the government of Mongolia has also not exhibited a willingness to enforce existing environmental laws that would deter or hold accountable those who egregiously pollute Mongolia's water sources and pastureland.

It is only fair to note that Chinese firms are not the only investors pursuing Mongolia's mineral wealth; Australian, Canadian, and US firms are also well invested. Russia too is setting its sights on Mongolia's minerals: Russian Railways (RZD) plans to invest 1.5 billion (US) dollars to modernize Mongolia's rail system mainly in order to expedite the export of Tavan Tolgoi coal into Russia.[17]

In spite of scattered cases of antiforeigner violence in Ulaanbaatar, xenophobia is not widespread. Yet, there is a perceptible fear that the Chinese are buying up Mongolia's assets, both in Ulaanbaatar and in the countryside. In 2010, however, the Mongolian government made a shrewd move in its decision to retain 100 percent state ownership of Tavan Tolgoi rather than auction off 49 percent of Tavan Tolgoi's coal to a foreign bidder.[18]

Thus far, it seems that herders have not had the time, organization, or political will to assemble an effective lobby for protecting pasturelands and water sources from the effects of mining by Mongolian, Chinese, and other foreign companies. The sagging confidence of herders in the political system is undeniable. An October 2010 survey by the Sant Maral Foundation found that only 29 percent of people in the countryside felt that "political parties represent public opinion" and 60.4 percent felt that political parties did not reflect public opinion.[19] The Sant Maral Foundation's May 2011 survey detected even lower confidence in the democratic process: only 22.7 percent of people in the countryside agreed that "political parties represent public opinion," with 64.3 percent feeling that the parties did not.[20] Simply put, herders feel their voices are not being heard. Thus, it may become the job of environmental watchdog groups like the Ongi River Movement to join forces with those elected officials who understand the dangers of signing over land use rights to foreign concerns.

Mongolia's economists, one hopes, see the need for a balanced GDP that includes the output of herders as well as the output of mining. Environmental advocates, herders, politicians, and economists may eventually see common cause in promoting the livestock herding sector and the pastoral nomadic way of life, though this outcome is not by any means guaranteed.

THE ROLE OF GOVERNMENT IN LAND USE: THE DRAFT LAW ON PASTURELAND

Recent scholarship has stressed the "institutional vacuum" in the Mongolian countryside in the postsocialist period.[21] In comparison with the role of the state in the socialist period, it is obvious that in the two decades since the collapse of socialism the Mongolian state has been far less involved in pasture allocation and pasture conflict resolution. This retreat from institutional responsibility, while understandable because of financial constraints over the past two decades, has had deep repercussions. With abiding uncertainty about possible future privatization of pastures, herders in the 1990s were building cabins on summer pastures in order to lay claim to good grazing lands.[22]

There are, however, indications that the Mongolian government is preparing to take on a more active role in pasture-use issues. In the fall 2010 and spring 2011 sessions of parliament, the Draft Law on Pastureland came under discussion. The current patchwork allocation of winter/spring pastures and livestock enclosures ideally should be regularized with one system expanded with as much uniformity as local and regional geographic conditions allow. Enforcement of currently existing rangeland boundaries has been very uneven, both for summer/autumn grazing and for winter/spring grazing. The pending pastureland law will address a variety of issues including pasture demarcation in *otor* reserve areas that are often inter-sum and inter-aimag; the regulation and restriction of livestock numbers based on winter/spring pastureland carrying capacity; uniform granting of certificates of possession to those who pay pasture-use fees; and settlement of pasture-related disputes.

Throughout the pending Draft Law on Pastureland, provision after provision empowers sum and aimag governors and citizens' representative *khurals* to regulate the leasing of pastureland and the resolving of pastureland disputes. Clearly, the government's role is envisioned as central to the implementing and overseeing of the numerous articles pertaining to land use. The sum governor may operate in conjunction with self-governing herder associations termed "Pasture User Groups" (PUGs). Article 8.3 in Chapter Two of the Draft Law on Pastureland attempts to clarify the relationship between top-down authority (sum governors) and PUG members:

> Soum governor and Soum Association of PUGs shall jointly form Soum Pasture Comanagement Committee consist[ing] of representatives of herder self-governing organization of soum, state inspector of environmental control, land manager and agricultural specialist.[23]

PUGs, thus, would be invested with the authority to develop mid-term and annual pasture management plans that would include determining appropriate pasture carrying capacity as well as drawing up a schedule of pasture rotation and livestock-grazing patterns. Thus, the PUGs would work in conjunction with local officials to regulate overall land use issues and to regulate herd numbers. Observers of PUGs in Arkhangai Aimag have noted that the focus on cooperative planning and pooled labor somewhat resembles that of the socialist-era *negdel* work teams, though the composition and functioning of the PUGs are far less hierarchical and more egalitarian than their socialist predecessors.[24]

If the government develops the legal framework and enforcement capacity at the local level to regulate pastureland usage, then regulating herd size by region and locale would help alleviate pressure on the land. Continual reevaluation would be necessary to measure the carrying capacity of an ever-changing landscape that is in the process of being locked up by mineral rights leases, expanding tourist bases, and other competing interests. In spring 2011, the *Ikh Khural* was in the process of collecting information on herder opinions on the Draft Law on Pastureland. This piece of legislation has the potential to balance top-down authority and bottom-up input, thereby achieving the best possible outcomes in environmental controls, economic development, and cultural sustainability.

While this book has devoted less attention to the so-called peri-urban zones, where city and countryside overlap in a jumble of outlying ger districts, there is an urgent need to address land use and property-rights issues in these zones. As more and more herding households are drawn to Ulaanbaatar, Darkhan, and Erdenet, Mongolia's three major cities, in order to seek schooling for their children and easier access to markets to sell their dairy, meat, and wool products, peri-urban regions have been subject to greater environmental damage than other areas. It will be essential to sort out pasture rights in order to avoid further overgrazing and pollution of water sources.

Mongolia's Millennium Challenge Account (MCA) is funding a so-called Peri-Urban Range Land Project that will introduce leasing of peri-urban pastureland to herders in outlying zones of Ulaanbaatar, Darkhan, and Erdenet.[25] Funding for new wells, fencing, and livestock shelters is earmarked, as selected herding groups will then receive leases in return for land use fees. Before this project can be enacted, however, the Draft Law on Pastureland that has been under discussion will, in some form, need to be passed by the Mongolian parliament to provide legal groundwork.

Much work remains to be done by Mongolia's parliament and relevant government ministries to guarantee that any new laws on pastureland are introduced in a fair, equitable manner. The inequitable redistribution of negdel holdings in the early 1990s led to an immediate gap between the wealthier herding households and the poorer ones. In order to avoid a repeat of this scenario, the enactment of laws and regulations pertaining to pasture leasing will need to be carried out in a transparent and inclusive manner with wide participation by herder groups.

The enormous mineral deposits that await development have led to the coining of the phrase "Mine-golia."[26] On the domestic front, mineral wealth will lead to social changes across all economic strata. For instance, the development of the Tavan Tolgoi coal fields that have an estimated 6.4 billion tons of coal reserves is expected to include a distribution of shares to every Mongolian citizen.[27] Along with equitable pasture leasing, the distribution of mining shares and mining revenues will open the door to potentially opaque and illicit dealings (as in the redistribution of state-owned assets in the early 1990s), with more savvy investors taking advantage of less knowledgeable shareholders. As in pastureland management, so also in mining revenue distribution, the state's regulatory authority will be crucial in determining equitable results.

On the international front, Mongolia's longtime perceived weakness as a landlocked nation is now evolving into a perceived strength as a mineral-rich "land-linked" country.[28] China and Russia are eyeing Mongolia's mineral deposits and are eagerly vying to build new railway lines to cart Mongolia's riches off to their respective markets. This revived "Great Game" between China and Russia is reminiscent of the early twentieth-century Russo-Japanese struggle over railroad dominion in North Asia. Mongolia's elected officials will face enormous temptations to garner personal profits from dealings with international mineral resource development partners, just as in the domestic arena politicians will undoubtedly find it difficult to balance the needs of quieter constituents like herders with those of more vocal and wealthier constituents who stand to profit from mining.

There is a divided opinion about the economic feasibility of pastoral nomadic herding in Mongolia. Thomas Barfield, an anthropologist with expertise in comparative nomadic cultures, is among those who are less sanguine about the ability of Mongolia's herders to thrive financially, but he voices the opinion that there is more to the picture than just economics:

> Still, the pastoral nomadic way of life has always kept itself distinct from the outside world, even as it interacted with it. As long as people have the means to preserve it as a way of life, the nomadic pastoral ethnoscape will endure.[29]

Indeed, it is crucial to keep in mind when predicting the future of pastoral herding in Mongolia that economics and culture (the pastoral nomadic way of life) may be weighted by herders differently from the way economists weigh the two. Ultimately, as Mongolia continues to develop its democratic institutions, herders themselves will decide on the value and merits of their way of life as economic temptations from other sectors may offer more lucrative paths away from herding. The land itself, however, has a long history of supporting pastoral nomadic herding.

Flexibility and openness to change are qualities that Mongolia's herders have demonstrated in the past and bode well for the continuation of this way of life. Those who express pessimism about the future of traditional herding in Mongolia underrate nomadic resiliency. Political upheavals, long periods of "statelessness," economic setbacks, and periodic ecological disasters have long been integral parts of nomadic history, both in Mongolia and in the greater Central Eurasian steppe. Risk and resiliency are the two sides of the equation in pastoral nomadism, as Mongolia's herders well know.

Glossary of Mongolian Terms

aimag	The largest administrative unit in Mongolia, similar to a province.
argal	Dried dung from livestock used by herders as fuel in woodstoves in *gers*.
ayil	An encampment of gers.
bag	The smallest administrative unit in Mongolia; a district.
ezemshigchiin gerchilgee	Government-issued certificates of possession for winter campsites and associated pastureland.
ger	The felt tent used by Mongolia's nomadic herders.
khot ayil	Nomadic encampment of gers; a social unit not necessarily based on kinship.
khural	An assembly. For instance, the *Ikh Khural* is Mongolia's parliament.
negdel	Socialist collective for herders.
otor	Long distance migration usually by a few members of a herding household with part of the livestock herd, often undertaken as a strategy to fatten up the livestock in autumn or to avoid drought or *zud*.
ovoo	Usually a cairn of stones that may mark a spiritual place such as a mountain pass; may also mark pasture boundaries.
sum	Subdivision of an aimag, similar to a county; larger than a bag.

tögrög	Mongolian unit of currency, in the socialist period artificially pegged at 4 *MNT* to the dollar; in early 2012, about 1300 *MNT* to US$1.
zag	Mongolian for saksaul, an endangered woody shrub that grows in the Gobi.
zud	Disasters caused by heavy snows or icing of pastureland that prevent livestock from accessing forage or caused by summer drought that stunts pastureland.

NOTES

1 Introduction

1. "Livestock Faces Dramatic Increase," *MWN,* December 25, 2006. The statistic of 34.8 million appeared in "Economic and Social Situation of Mongolia," *MM* 3:809 (January 17, 2007), 6.
2. Quoted speech is from B. Ooluun, "New Year's Eve at Sukhbaatar Square," *MM* 1:807 (January 3, 2007), 1.
3. See Mongolian Society for Range Management, *Livelihood Study of Herders in Mongolia* (Swiss Agency for Development and Cooperation, June 2010), xx, available on the SDC website, www.swiss-cooperation.admin.ch/ . . . resource_en_184798.pdf.
4. For a statistical ranking of replies concerning livestock priorities, see Open Society Forum, *The Future of Nomadic Pastoralism in Mongolia: Public Perception Survey* (Ulaanbaatar: Open Society Forum, 2004), 89–90.
5. Linda Gardelle, *Pasteurs nomades de Mongolie: Des sociétés nomades et des États* (Paris : Buchet/ Chastel, Écologie, 2010), 37–38.
6. See Wietze Lise, Sebastiaan Hess, and Byamba Purev, "Pastureland Degradation and Poverty among Herders in Mongolia: Data Analysis and Game Estimation," *Ecological Economics* 58 (2006): 350–364.
7. Gardelle, *Pasteurs nomades,* 42
8. *MM* 1:1018 (January 7, 2011), 3.
9. "Social and Economic Situation of Mongolia (as of the Preliminary Result of 2011)," *MM* 4:1073 (January 27, 2012), 6.
10. See Swiss Agency for Development and Cooperation in Mongolia, *Annual Report 2007,* 4, available at the SDC website, www.sdc.mn
11. Troy Sternberg cites Mongolia's 2009 Statistical Yearbook for this figure. See Sternberg, "Unravelling Mongolia's Extreme Winter Disaster of 2010," *Nomadic Peoples* 14, no. 1 (2010): 79.

12. Swiss Agency for Development and Cooperation in Mongolia, *Annual Report 2009*, 3, available on the SDC website, www.sdc.mn
13. Dennis Ojima and Togtohyn Chuluun, "Policy Changes in Mongolia: Implications for Land Use and Landscapes," in *Fragmentation in Semi-Arid and Arid Landscapes: Consequences for Human and Natural Systems*, eds. K. A. Galvin, R. S. Reid, R. H. Behnke, Jr., and N. T. Hobbs (Dordrecht, The Netherlands: Springer, 2008), 189.
14. Definition provided by the Mongolian Society for Range Management, *Livelihood Study*, xiii.
15. A. D. Simukov, "O kochevkakh i pastbishchakh Mongolii," originally published in 1929 in the journal *Khoziaistvo Mongolii* 2 no. 158: 5–24; republished in *A. D. Simukov, Trudy o Mongolii i dlia Mongolii*, eds. Yuki Konagaya, Sanzhaasurengiin Bayaraa, and Ichinkhorloogiin Lkhagvasuren, vol. 2 (Osaka: Gosudarstvennyi muzei etnologii, 2007), 358–376. Quote from page 358.
16. Simukov, "O kochevkakh i pastbishchakh Mongolii," 360.
17. See the press release, "Mongolia's Pasture Quality in Worrying Decline," *MWN*, July 29, 2009.
18. Saverio Krätli, "Editorial," *Nomadic Peoples* 14, no. 1 (2010): 4.
19. B. Bulgamaa, "Disasters Bring Mongolian Herders to Their Knees," *UBP*, January 15, 2009.
20. See Arshad Sayed, "2010 Dzud Is Not Entirely a Disaster of Nature," *MM* 05:970 (February 5, 2010,) 7.
21. Sternberg, "Unravelling Mongolia's Extreme Winter," 81.
22. Sternberg, "Unravelling Mongolia's Extreme Winter," 81.
23. See David Sneath, "The Rural and the Urban in Pastoral Mongolia," in *Mongols from Country to City: Floating Boundaries, Pastoralism and City Life in the Mongol Lands*, eds. Ole Bruun and Li Narangoa (Copenhagen: Nordic Institute of Asian Studies, 2006), 151.
24. Dulamsuren Dorligsuren, Batjav Batbuyan, Bulgamaa Densambu, and Steven R. Fassnacht, "Lessons from a Territory-Based Community Development Approach in Mongolia: Ikhtamir Pasture User Groups," in *Restoring Community Connections to the Land: Building Resilience through Community-Based Rangeland Management in China and Mongolia*, eds. Maria E. Fernandez-Gimenez, Xiaoyi Wang, B. Batkhishig, Julia A. Klein, and Robin S. Reid (Cambridge, MA: CAB International, 2012), 174.
25. V. V. Graivoronskii, *Ot kochevogo obraza zhizni k osedlosti (na optye MNR)* (Moscow: "Nauka", 1979), 3 and 153.
26. Graivoronskii, *Ot kochevogo obraza zhizni*, 154.
27. V. V. Graivoronskii, *Reformy v sotsial'noi sfere sovremennoi Mongolii* (Moscow: Institut vostokovedeniia RAN, 2007), 199.

28. Ole Bruun, *Precious Steppe: Mongolian Nomadic Pastoralists in Pursuit of the Market* (New York: Lexington Books, 2006), 236.
29. Sternberg, "Unravelling Mongolia's Extreme Winter," 76–78.
30. See A. J. Dietz, E. Amgalan, T. Erdenechuluun, and S. Hess, "Carrying Capacity Dynamics, Livestock Commercialization and Land Degradation in Mongolia's Free Market Era," Poverty Reduction and Environmental Management (PREM) Working Paper (June 2, 2005), 1–15, http://www.prem-online.org/archive/2/doc/PREM%20WP%2005-10.pdf.
31. Kris Havstad, Jeffrey Herrick, and Enkh-Amgalan Tseelei, ""Mongolia's Rangelands: Is Livestock Production the Key to the Future?" *Frontiers in Ecology and the Environment* 6 (September 1, 2008): 389.
32. Mongolian Society for Range Management, *Livelihood Study*, 1, 3.
33. Mongolian Society for Range Management, *Livestock Study*, Chapter 2.
34. Gardelle, *Pasteurs nomades*, 163–164.
35. See Emanuel Marx, "Are There Pastoral Nomads in the Arab Middle East?" in *The Anthropology of Tribal and Peasant Pastoral Societies: The Dialectics of Social Cohesion and Fragmentation*, eds. Ugo Fabietti and Philip Carl Salzman (Pavia: Collegio Ghislieri, 1996), 101–115; see especially page 108.
36. Joseph Ginat, "Preservation and Change in Bedouin Societies in Israel," in *Changing Nomads in a Changing World*, eds. Joseph Ginat and Anatoly M. Khazanov (Portland, OR: Sussex Academic Press, 1998),171–201; quote on 171–172.
37. Kenneth M. Bauer, "Common Property and Power: Insights from a Spatial Analysis of Historical and Contemporary Pasture Boundaries among Pastoralists in Central Tibet," *Journal of Political Ecology* 13 (2006): 24.
38. See G. Chingis, "Tavan Tolgoi—A Great Project or a Huge Joke?" *UBP*, July 20, 2010; and G. Chingis, "Are We Russian or Mongolian?" *UBP*, July 27, 2010.
39. See Open Society Forum, *The Future of Nomadic Pastoralism*, 41.
40. Gardelle, *Pasteurs nomades*, 78–80.
41. John Bell, *A Journey from St. Petersburg to Pekin 1719–22* (1763; repr., New York: Barnes & Noble, 1966), 89.
42. Roy Chapman Andrews, *Across Mongolian Plains: A Naturalist's Account of China's "Great Northwest"* (New York: Blue Ribbon Books, 1921), 78.
43. "Mongolians Gather to Protest Rising Food Prices," *MWN*, April 21, 2008.

44. See, for instance, "Spring Crop Details Announced by Officials," *MWN,* May 15, 2008.
45. See Mongolian Society for Range Management, *Livelihood Study,* 5.
46. "Data Shows 28% of Mongolia Set Aside for Mining and Exploration," *MWN,* July 25, 2008; see also the Open Society Forum's report, "License Watch #5: Status of Awards of Licenses in Mining as of 9 May 2008" (2008), www.opensocietyforum.mn.
47. Open Society Forum, "License Watch #5," 9–10.
48. On China's economic involvement in Africa, see Howard W. French, "The Next Empire?" *The Atlantic* (May 2010): 58–69.
49. See "Not Enough Qualified Locals, Minister Wants to Import Chinese Labor for OT," *MWN,* June 7, 2010.
50. "More Mongolians Employed at Oyu Tolgoi Than Stipulated, Says Minister," *MWN,* July 14, 2010.
51. P. Shinebayar, "1,000 Young People to Get Skills Beginning Sep 1," *UBP,* August 6, 2010.
52. P. Shinebayar, "Immigrant Workers Enter Oyu Tolgoi in Storm of Dissent," *UBP,* May 20, 2011.
53. H. G. C. Perry-Ayscough and Captain R. B. Barry, *With the Russians in Mongolia* (New York: John Lane Co., 1914), 167–168.
54. Perry-Ayscough and Barry, *With the Russians in Mongolia,* 5.
55. Perry-Ayscough and Barry, *With the Russians in Mongolia,* 4.
56. Yuki Konagaya and I. Lkhagvasuren, "Tsogt-Ochiriin Lookhuuz: Former Director of State Farms," in *Socialist Devotees and Dissenters: Three Twentieth-Century Mongolian Leaders,* ed. Morris Rossabi, trans. Mary Rossabi (Osaka: National Museum of Ethnology, 2011), 88.
57. Data is from "Eighth Semi-Annual Corruption Benchmarking Survey", *MM* 52:964 (December 23, 2009), 6–7.
58. "Businessmen Vocal about Having to 'Pay Bribes at Every Step,'" *MWN,* October 1, 2010.
59. See "Businessmen Vocal about Having to 'Pay Bribes at Every Step'"; and "Human Rights Commission Finds Questions for Trainees Violate Law," *MWN,* October 1, 2010.
60. Statistics are from the SDC, *Annual Report 2009,* 22, available on the SDC website, www.sdc.mn

2 Mongolia's Regions: Lessons in Variation

1. See J. M Suttie, "Grazing Management in Mongolia," in *Grasslands of the World,* eds. J. M. Suttie, S. G. Reynolds, and C. Batello

(Rome: Food and Agriculture Organization of the United Nations, 2005), 277–278.

2. Information on the Gobi is drawn from the following sources: Bernd Steinhauer-Burkart, *Gobi Gurvansaikhan National Park* (Ulaanbaatar: Eco-Nature, 1999); C. Finch, *Mongolia's Wild Heritage: Biological Diversity, Protected Areas, and Conservation in the Land of Chingis Khaan* (Ulaanbaatar: Mongolia Ministry for Nature and Environment, 1996); and Christopher P. Atwood. "Gobi Desert," *EMME* (New York: Facts on File, 2004), 200–201.
3. On Gobi wells and water access issues, see David Lazar, "Well Water Pumps Provide New Reliability and Water Supply for Mongolian Sheepherders," *Water Conditioning & Purification Magazine* (September 2004): 8–9; and "Impacts of Well Rehabilitation and Human Intrusion on Khulan (Wild Ass) and Other Threatened Species in the Gobi Desert," http:khulan.autonomyproductions .com/files/IntroMethods.pdf. (2009).
4. See B. Bulgamaa, "31 percent of Mongolian Territory Has Been Issued Mining Licenses," *UBP,* February 5, 2009.
5. See "Ivanhoe Mines and Rio Tinto Sign Long-Term Investment Agreement with Mongolia to Build and Operate Oyu Tolgoi Copper-Gold Mining Complex," (October 6, 2009), http://news.moneycentral.msn.com.
6. See B. Ooluun, "Mongolians to Share TT Profits," *MM* 23:988 (June 11, 2010), 1, 3; and "More Mongolians Employed at Oyu Tolgoi than Stipulated, Says Minister," *MWN,* July 14, 2010.
7. See Sh. Batmonkh, "MPs Suggest Inflating Cashmere Prices," *The UBP,* May 22, 2008.
8. Information on the life of A. D. Simukov is drawn from his daughter, N. A. Simukova's, account, "Andrei Dmitrievich Simukov, zhizn' i trudy," in *A. D. Simukov: Trudy o Mongolii i dliia Mongolii,* eds. Yuki Konagaya, Sanzhaasurengiin Bayaraa, and Ichinkhorloogiin Lkhagvasuren, *A. D. Simukov: Trudy o Mongolii i dliia Mongolii,* vol. 1 (Osaka: Gosudarstvennyi Muzei Etnologii, 2007), 15–29.
9. This work, entitled in Russian "Gobiiskie pastbishcha Mongol'skoi Narodnoi Respubliki," is republished in Konagaya et al., eds., *A. D. Simukov,* vol. 2, 5–202. The original piece was 208 pages in typeset; it is a bit shorter as reprinted in the 2007 volume. See the list of Simukov's works in vol. 3, part 2 of Konagaya et al., *A. D. Simukov,* 231.
10. Linda Gardelle, *Pasteurs nomades de Mongolie: Des sociétés nomades et des États* (Paris: Buchet/Chastel, Écologie, 2010), 111–112.

11. Natasha Fijn, *Living with Herds: Human-Animal Coexistence in Mongolia* (New York: Cambridge University Press, 2011), 183.
12. These and other observations by Simukov in this section are drawn from Simukov, "O kochevkakh i pastbishchakh Mongolii," originally published in 1929 in the journal *Khoziaistvo Mongolii* 2 no. 158: 5–24; republished in *A. D. Simukov, Trudy o Mongolii i dlia Mongolii*, eds. Yuki Konagaya, Sanzhaasurengiin Bayaraa, and Ichinkhorloogiin Lkhagvasuren, vol. 2 (Osaka: Gosudarstvennyi Muzei Etnologii, 2007) 371–373, unless noted otherwise.
13. Simukov, "O kochevkakh i pastbishchakh Mongolii," 373.
14. Simukov, "Gobiiskie pastbishcha Mongol'skoi Narodnoi Respubliki," 132. The following data and observations are drawn from the same work by Simukov.
15. Simukov, "Gobiiskie pastbishcha," 14.
16. Simukov, "O kochevkakh i pastbishchakh Mongolii," 374.
17. Simukov, "Gobiiskie pastbishcha," 122.
18. B. Oyundelger, "In 20 Years, There May Be No Camels," *MM* 42:1007 (October 22, 2010), 9.
19. See J. M. Suttie, "Grazing Management in Mongolia," 286–287.
20. Simukov, "O kochevkakh i pastbishchakh Mongolii," 371.
21. Quote from page 364 of Simukov, "O kochevkakh i pastbishchakh Mongolii"; see also page 130 of Simukov, "Gobiiskie pastbishcha."
22. Simukov, "O kochevkakh i pastbishchakh Mongolii," 365.
23. J. M. Suttie, "Grazing Management in Mongolia," 279.
24. Fijn, *Living with Herds*, 187.
25. Simukov, "O kochevkakh i pastbishchakh Mongolii," 367.
26. Ole Bruun, *Precious Steppe: Mongolian Nomadic Pastoralists in Pursuit of the Market* (New York: Lexington Books, 2006), 56–57.
27. J. M Suttie, "Case Study 8. Hay from Natural Pasture in Mongolia (The Change from Cooperative to Private Stock-Rearing in a Purely Pastoral Economy)," in *Hay and Straw Conservation for Small-Scale Farming and Pastoral Conditions* by J. M. Suttie (Rome: Food and Agriculture Organization of the United Nations, 2000): 237–243.
28. Swiss Agency for Development and Cooperation, "'Green Gold' Pasture Ecosystem Management Project" (September 2009), www.swisscooperation.admin.ch/mongolia//. . . //resource_en_184877.pdf.
29. Suttie, "Grazing Management in Mongolia," 278.
30. Simukov, "O kochevkakh i pastbishchakh Mongolii," 368.
31. Suttie, "Case Study 8."

32. Suttie, "Grazing Management in Mongolia," 279.
33. See the project summary at "Gobi Forage," http://www.usaid.gov/mn/programs/gobiforage/index.html.
34. Myatavyn Tumurjav, "Traditional Animal Husbandry Techniques Practiced by Mongolian Nomadic People," in *Mongolia Today: Science, Culture, Environment and Development*, eds. Dendeviin Badarch, Raymond A. Zilinskas, and Peter J. Balint (New York: RoutledgeCurzon, 2003), 106.
35. Simukov, "Gobiiskie pastbishcha," 144.
36. On modern-day conservation of the saksaul, see Steinhauer-Burkart, "Gobi Gurvansaikhan National Park," 52–53.
37. Nicholas Crisp, John Dick, and Michael Mullins, *Mongolia: Forestry Sector Review* (Victoria BC: The World Bank, April 2004), 107, http://siteresources.worldbank.org/INTMONGOLIA/Resources/eng_version.pdf.
38. Crisp, et al., *Mongolia: Forestry Sector Review*, 106–107.
39. J. Tsogtbaatar, "Deforestation and Reforestation Needs in Mongolia," *Forest Ecology and Management* 201 (2004), 106–107.
40. Simukov, "O kochevkakh i pastbishchakh," 362.
41. See Simukov, "Mongol'skie kochevki," in Konagaya et al., *A. D. Simukov: Trudy o Mongolii i dliia Mongolii*, vol. 2, 443–449. Information on Simukov's six basic types of nomadic migration is drawn from this article.
42. Sandra L. Olsen, "Nomadic Pastoralists, Their Livestock, and Their Landscape," www.carnegiemnh.org/anthro/olsen_mongolia1.html.
43. Fijn, *Living with Herds*, 185.
44. Jargalmaa Gombo-Ochir, "Funds to Build Mongolia's Dairy Industry," *MM* 40:1005 (October 8, 2010), 8.
45. B. Ooluun, "Cashmere Products Become Luxury Items; State Urged to Ban Export of Raw Material," *MM* 3:1020 (January 21, 2011), 5.
46. Sh. Erdenechimeg, "All Mongolian Livestock to Be Registered," *MM* 21:1038 (May 27, 2011), 1, 3.
47. Suttie, "Grazing Management in Mongolia," 298–299.

3 Early History

1. See Bat-Ochir Bold, *Mongolian Nomadic Society: A Reconstruction of the "Medieval" History of Mongolia* (New York: St. Martin's Press, 2001), 57.

2. *The Secret History of the Mongols; A Mongolian Epic Chronicle of the Thirteenth Century*, vol. 1, trans. Igor de Rachewiltz (Leiden and Boston: Brill, 2004), 149, 150.
3. See *Secret History* vol. 1, 214 and vol. 2, 1026.
4. *Secret History*, vol. 1, 217. I have not retained de Rachewiltz's italics here.
5. *Secret History*, vol. 1, 218. The literature on the economic, political, and military dimensions of hunting among the imperial era Mongols is extensive. See, for a start, Sechin Jagchid and Paul Hyer, *Mongolia's Culture and Society* (Boulder, CO: Westview Press, 1979), 27–37. De Rachewiltz lists several works on Mongolian hunting practices in volume 1, page 436 of his *Secret History.*
6. *Secret History*, vol. 1, 2. See also de Rachewiltz's comments on this passage in volume 1, page 248.
7. Mark D. Johnstad and Richard P. Reading, "Mongolia's Protected Areas System," *Biodiversity Briefings from Northern Eurasia* 2, no. 1 (2003): 8.
8. *Secret History*, vol. 1, 127. See also de Rachewiltz's commentary on this passage in volume 2, page 738.
9. Thomas T. Allsen, *The Royal Hunt in Eurasian History* (Philadelphia, PA: University of Pennsylvania Press, 2006), 96.
10. *Secret History* vol. 1, 192 and vol. 2, 950.
11. For evidence, see Valerie Hansen, *Negotiating Daily Life in Traditional China: How Ordinary People Used Contracts 600–1400* (New Haven, CT: Yale University Press, 1995), 115–120.
12. Elizabeth Endicott-West, *Mongolian Rule in China: Local Administration in the Yuan Dynasty* (Cambridge: Harvard University Press, 1989), 120.
13. Igor de Rachewiltz, "Yeh-lü Ch'u-ts'ai (1189–1243), Yeh-lü Chu (1221–1285)," in *In the Service of the Khan: Eminent Personalities of the Early Mongol-Yüan Period (1200–1300)*, eds. Igor de Rachewiltz, Hok-lam Chan, Hsiao Ch'i-ch'ing, and Peter W. Geier (Wiesbaden: Harrassowitz, 1993), 149.
14. Endicott-West, *Mongolian Rule*, 55–56.
15. This imperial decree is translated in Endicott-West, *Mongolian Rule*, 57.
16. See Elizabeth Endicott-West, "The Yüan Government and Society," in *Alien Regimes and Border States, 907–1368*, vol. 6 of *The Cambridge History of China*, eds. Herbert Franke and Denis Twitchett (New York: Cambridge University Press, 1994), 598–599.

17. The touxia system has been explored in depth in Endicott-West, *Mongolian Rule*, Chapter Four, 89–103.
18. See the I. J. Schmidt edition of the *Erdeni-yin tobči: Geschichte der Ost-Mongolen und ihres Fürstenhauses verfasst von Ssanang Ssetsen Chungtaidschi der Ordus* (Beijing 1937 reprint of the St. Petersburg, N. Gretsch, 1829 ed.), 138, lines 1–2.
19. For several examples of Yuan imperial decrees extending aid to herding groups in the Mongolian homeland, see Elizabeth Endicott, "The Mongols and China: Cultural Contacts and the Changing Nature of Pastoral Nomadism (Twelfth to Early Twentieth Centuries)," in *Mongols, Turks, and Others: Eurasian Nomads and the Sedentary World*, eds. Reuven Amitai and Michal Biran (Leiden: Brill, 2005), 467–472.
20. See Udo B. Barkmann, "Some Comments on the Consequences of the Decline of the Mongol Empire on the Social Development of the Mongols, " in *The Mongol Empire and Its Legacy*, eds. Reuven Amitai-Preiss and David O. Morgan (Leiden: Brill, 1999), 276–277.
21. See Henry Serruys, "Landgrants to the Mongols in China: 1400–1460," in Henry Serruys, *The Mongols and Ming China: Customs and History*, ed. Françoise Aubin (London: Variorum Reprints, 1987), 394–395. (Originally published in *Monumenta Serica* XXV, 1966).
22. For a succinct overview of Altan Khan's life, see Christopher P. Atwood, "Altan Khan," *EMME* (New York: Facts on File, 2004), 9–10.
23. See Henry Serruys, "Altan-qaγan," in *Dictionary of Ming Biography 1368–1644*, vol. 1, eds. L. Carrington Goodrich and Chaoying Fang (New York: Columbia University Press, 1976), 6–9.
24. See Henry Serruys, "Notes on a Chinese Inscription of 1606 in a Lamaist Temple in Mai-ta-chao, Suiyüan," *Journal of the American Oriental Society* 78, no. 2 (April–June, 1958):101–113.
25. See Robert J. Miller, "Areas and Institutions in Eastern Asia," *Southwestern Journal of Anthropology* 9, no. 2 (Summer, 1953): 204–208.
26. See Walther Heissig, *The Religions of Mongolia*, trans. Geoffrey Samuel (Berkeley, CA: University of California Press, 1980), Chapter 4.
27. On folk religion and its survival, see Heissig, *Religions of Mongolia*, Chapter 5.

28. M. Sanjdorj, *Manchu Chinese Colonial Rule in Northern Mongolia*, trans. and annotated by Urgunge Onon (New York: St. Martin's Press, 1980), 1.
29. Robert J. Miller, "Buddhist Monastic Economy: The Jisa Mechanism," *Comparative Studies in Society and History* 3, no. 4 (July, 1961): 435.
30. Miller, "Buddhist Monastic Economy," 436.
31. Sanjdorj, *Manchu Chinese Colonial Rule*, 1.
32. The following points are drawn from George Murphy, "Comment," *Comparative Studies in Society and History* 3 no. 4 (July, 1961): 439–442.
33. On this so-called Tibetan-script Mongolian, see Christopher P. Atwood, "Tibetan Language and Script," in *EMME,* 538.
34. See C. R. Bawden, *The Modern History of Mongolia* (New York: Praeger, 1968), 247–248.
35. For details, see György Kara, *Books of the Mongolian Nomads: More Than Eight Centuries of Writing Mongolian*, trans. John R. Krueger (Bloomington, IN: Indiana University Research Institute for Inner Asian Studies, 2005), 74–77.
36. Walther Heissig, *A Lost Civilization: The Mongols Rediscovered*, trans. D. J. S. Thomson (New York: Basic Books, 1966), 96–107.
37. For a full translation of this imperial rescript, see F. W. Cleaves, "A Mongolian Rescript of the Fifth Year of Degedü Erdem-tü (1640)," *Harvard Journal of Asiatic Studies* 46 (June, 1986): 186–187.
38. Bawden, *Modern History of Mongolia*, 80.
39. For a detailed description of the Inner Mongolian and Outer Mongolian banners, see Christopher P. Atwood, "Banners," in *EMME*, 30–32.
40. Atwood, "Banners," 30.
41. See Peter C. Perdue, *China Marches West: The Qing Conquest of Central Eurasia* (Cambridge: Harvard University Press, 2005), 175–176.
42. Sechin Jagchid and Paul Hyer, *Mongolia's Culture and Society* (Boulder, CO: Westview Press, 1979), 278.
43. O. Pürev, "Khalkhyn Setsenkhan aimgiin borzhigin setsen vangiin khoshuuny tüükhen gazarzüi," *Borzhigin sudlal* I (Ulaanbaatar, 1999), as cited in L. Altanzaia, "Khalkhyn Aimag Khoshuudyn nutgiin zakhyn tukhai asuudalk," *Tüükhiin Setgüül* (Ulaanbaatar, 2003), 84.
44. See Jagchid and Hyer, *Mongolia's Culture and Society*, 275 and 278.
45. Bat-Ochir Bold, *Mongolian Nomadic Society*, 47.

46. For a concise write-up of the spiritual functions of the ovoo, see Christopher P. Atwood, "*oboo*," in *EMME*, 414–415.
47. See Henry Serruys, "A Question of Land and Landmarks between the Banners Otoγ and Üüsin (Ordos), 1904–1906," *Zentralasiatische Studien* 13 (1979): 215–237.
48. C. R. Bawden, "A Joint Petition of Grievances Submitted to the Ministry of Justice of Autonomous Mongolia in 1919," *Bulletin of the School of Oriental and African Studies, University of London* 30, no. 3 (1967): 548.
49. For a summation of these negative consequences, see Bawden, "Joint Petition," 548–549.
50. The quotations are from the autobiography of Jamsrangiin Sambuu (1895–1972), a high-ranking Mongolian government official in the socialist era. See *Herdsman to Statesman. The Autobiography of Jamsrangiin Sambuu of Mongolia*, trans. Mary Rossabi (New York: Rowman & Littlefield, 2010), 50, 46.
51. Š. Rasidondug and Veronika Veit, *Petitions of Grievances Submitted by the People (18th—beginning of 20th century)* (Wiesbaden: Otto Harrassowitz, 1975), 27.
52. Rasidondug and Veit, *Petitions*, 99.
53. *Moskovskaia torgovaia ekspeditsiia v Mongoliiu* (Moscow: P.P. Riabushinskii Printing House, 1912), 99. This source will hereafter be abbreviated as *MTE*.
54. *MTE*, 116.
55. *MTE*, 117.
56. *MTE*, 157.
57. Jagchid, *Mongolia's Culture and Society*, 315–316.
58. See photos ##18, 19, 20 in Elizabeth Endicott, *Pages from the Past: The 1910 Moscow Trade Expedition to Mongolia* (Norwalk, CT: EastBridge, 2007).
59. See Christopher P. Atwood, "Farming," in *EMME*, 176.
60. See Timothy May, *The Mongol Art of War* (Yardley, PA: Westholme Publishing, 2007), 54–57 and 63.
61. For historical photos of camel caravans in Mongolia in the late Qing, see Martha Avery, *The Tea Road: China and Russia Meet across the Steppe* (Beijing: China Intercontinental Press, 2003), photos following page 96.

4 The Socialist Era

1. On this policy shift, see Mei-hua Lan, "China's 'New Administration' in Mongolia," in *Mongolia in the Twentieth Century: Landlocked*

Cosmopolitan, eds. Stephen Kotkin and Bruce A. Elleman (Armonk, NY: M. E. Sharpe, 1999), 39–58.

2. See M. Sanjdorj, *Manchu Chinese Colonial Rule in Northern Mongolia,* trans. from the Mongolian and annotated by Urgunge Onon (New York: St. Martin's Press, 1980), 32–33, on the Qing permit system; and see S. C. M. Paine, *Imperial Rivals: China, Russia, and Their Disputed Frontier* (Armonk, NY: M. E. Sharpe, 1996), 279, on Mongolian debt.
3. A variety of sources spin the events of 1911–1921 in different ways. See, for instance, C. R. Bawden, *The Modern History of Mongolia* (New York: Praeger, 1968), 187–205; Liu Xiaoyuan, *Reins of Liberation: An Entangled History of Mongolian Independence, Chinese Territoriality, and Great Power Hegemony, 1911–1950* (Stanford, CA: Stanford University Press, 2006), 6–9; and S. C. M. Paine, *Imperial Rivals*, 287–342.
4. See Christopher P. Atwood, "1921 Revolution," in *EMME* (New York: Facts on File, 2004), 471–476 for an overview of events.
5. Atwood, "1921 Revolution," 474.
6. Atwood, "New Turn Policy," in *EMME*, 405–406.
7. For an in-depth discussion of collectivization and secularization in the USSR, see Christopher Marsh, *Religion and the State in Russia and China: Suppression, Survival, and Revival* (London: Continuum International Publishing Group, 2011), 60–69.
8. Atwood, "New Turn Policy," 405.
9. See Nasan Dashdendeviin Bumaa, "The Twentieth Century: From Domination to Democracy," in *Modern Mongolia: Reclaiming Genghis Khan*, ed. Paula L. Sabloff (Philadelphia, PA: University of Pennsylvania Museum of Archaeology and Anthropology, 2001), 46–47.
10. The Academy of Sciences of the MPR, *Information Mongolia* (New York: Pergamon Press, 1990), 55.
11. Jamsrangiin Sambuu, *Herdsman to Statesman. The Autobiography of Jamsrangiin Sambuu of Mongolia*, trans. from the Mongolian by Mary Rossabi (New York: Rowman & Littlefield, 2010), 115.
12. Information Mongolia, 127.
13. Bumaa, "The Twentieth Century," 51.
14. Bawden, *Modern History*, 399.
15. Bawden, *Modern History*, 400–401.
16. *Information Mongolia*, 128.
17. Ts. Namkhainyambuu, *Bounty from the Sheep: Autobiography of a Herdsman*, trans. Mary Rossabi (Cambridge, UK: The White Horse Press, 2000), 57.

18. Namkhainyambuu, *Bounty*, 89.
19. Sh. Natsagdorj, "The Economic Basis of Feudalism in Mongolia," trans. Owen Lattimore, *Modern Asian Studies* 1, no. 3 (1967): 267, 281.
20. Bumaa, "The Twentieth Century," 54.
21. See, for instance, Linda Gardelle's comments in *Pasteurs nomades de Mongolie: Des sociétés nomades et des États* (Paris: Buchet/Chastel, Écologie, 2010), 50.
22. See "Appendix: A Manual for Herders," 103–150 in Namkhainyambuu, *Bounty from the Sheep.*
23. Gardelle elaborates upon this point in her *Pasteurs nomades*, 34–35.
24. Fijn, *Living with Herds*, 26.
25. Gardelle, *Pasteurs nomades*, 48.
26. Natasha Fijn, *Living with Herds: Human-Animal Coexistence in Mongolia* (New York: Cambridge University Press, 2011), 30.
27. Fijn, *Living with Herds*, 63.
28. Atwood, "Collectivization and Collective Herding," in *EMME*, 115–116.
29. Namkhainyambuu, *Bounty*, 77.
30. Namkhainyambuu, *Bounty*, 84.
31. Namkhainyambuu, *Bounty*, 72.
32. Namkhainyambuu, *Bounty*, 72.
33. Gardelle, *Pasteurs nomades*, 53.
34. V. V. Graivoronskii, *Kooperirovannoe aratstvo MNR: izmeneniia v urovne zhizni 1960–1980* (Moscow: "Nauka," 1982), 164.
35. Gardelle, *Pasteurs nomades*, 52–53.
36. Graivoronskii, *Kooperirovannoe aratstvo MNR*, 164–165.
37. Morten Axel Pedersen, *Not Quite Shamans: Spirit Worlds and Political Lives in Northern Mongolia* (Ithaca, NY: Cornell University Press, 2011), 135.
38. Pedersen, *Not Quite Shamans*, 142–143, n. 23.
39. On Russian-Mongolian relations, see Tsedendamba Batbayar, "Mongolian-Russian Relations in the Past Decade," *Asian Survey* 43, no. 6 (November/December, 2003): 951–970.
40. Melvyn C. Goldstein and Cynthia M. Beall, *The Changing World of Mongolia's Nomads* (Berkeley, CA: University of California Press, 1994), 76.
41. Namkhainyambuu, *Bounty*, 84.
42. Namkhainyambuu, *Bounty*, 89.
43. Namkhainyambuu, *Bounty*, 75.

44. Goldstein and Beall, *The Changing Nature of Mongolia's Nomads*, 80–83.
45. "Mongolia Human Development Report 2011," 50, citing National Statistical Office, 2011.
46. See "Russia, Mongolia Settle Mongolian Debts—Putin," *MM* 50, 51:1015, 1016 (December 17, 2010), 5.
47. See "Herders Awarded for Best Practices," *MM* 234:989 (June 18, 2010).
48. G. Chingis, "Are We Russian or Mongolian?" *UBP*, July 27, 2010.
49. Robert Rupen, How Mongolia Is Really Ruled: A Political History of the Mongolian People's Republic 1900–1978 (Stanford, CA: Stanford University Hoover Institution Press, 1979), 108.
50. G. Chingis, "Tavan Tolgoi—A Great Project or a Huge Joke?," *UBP*, July 20, 2010.
51. David Sneath, "Lost in the Post: Technologies of Imagination, and the Soviet Legacy in Post-Socialist Mongolia," *Inner Asia* 5, no. 1 (2003): 49.

5 Post-1990 Developments: Who Determines Land Use in the Era of the Free Market?

1. On the disbanding of negdels and decollectivization in the countryside, see Christopher P. Atwood, "Decollectivization" (pages 140–141) and "Privatization" (pages 440–441) in *EMME*.
2. Melvyn C. Goldstein and Cynthia M. Beall, *The Changing World of Mongolia's Nomads* (Berkeley, CA: University of California Press, 1994), 128.
3. Goldstein and Beall, *Changing World*, 160.
4. See David Sneath, "Land Use, the Environment and Development in Post-Socialist Mongolia," *Oxford Development Studies* 31, no. 4 (December 2003): 455, n. 7.
5. Robin Mearns, "Decentralisation, Rural Livelihoods and Pasture-Land Management in Post-Socialist Mongolia," *European Journal of Development Research* 16, n. 1 (Spring 2004): 140.
6. Again, see Mearns, "Decentralisation," 140, on the unfulfilled potential of the *khot ail* in the early 1990s.
7. Jeremy Swift, "Rural Development: The Livestock Sector," in *Poverty and the Transition to a Market Economy in Mongolia*, ed. Keith Griffin (New York: St. Martin's Press, 1995), 105.
8. Mearns, "Decentralisation," 140; see also the example in Maria E. Fernandez-Gimenez, "Spatial and Social Boundaries and the

Paradox of Pastoral Land Tenure: A Case Study from Postsocialist Mongolia," *Human Ecology: An Interdisciplinary Journal* 30, no. 1 (March 2002): 67, n. 3.

9. Martha Avery, *Women of Mongolia* (Seattle, WA: University of Washington Press, 1996), 112.
10. Avery, *Women of Mongolia*, 112.
11. Sneath, "Land Use," 443.
12. Sneath, "Land Use," 443–444.
13. See Mearns, "Decentralisation," 142–144.
14. UNDP, "Human Development Report: Mongolia 2003," 40.
15. Mearns, "Decentralisation," 144, notes the inherent contradictions between the 2002 legislation and earlier land law provisions.
16. See The World Bank, "Mongolia: Environment Monitor 2004," http://www-wds.worldbank.org, 22–24, on urban land markets.
17. See B. Altantuul, "Heated Exchanges over Land at Ikh Khural," *MTW,* June 1–3, 2002.
18. See B. Altantuul, "New Draft Tests Dangerous Waters," *MTW,* June 11, 2002.
19. "PM Defends Land Privatization," *MTW,* July 2, 2002.
20. The World Bank, "Mongolia: Environment Monitor 2004," 23.
21. See, for instance, "The Last Best Place," *The Economist* (December 21, 2002): 60; and Sneath, "Land Use," 449.
22. See "The Last Best Place," 60.
23. "No More Nomads in Mongolia, Prime Minister Says," *Far Eastern Economic Review* (May 23, 2001), http://www.feer.com/breaking_news/010523a.html.
24. "Mongolian Nomadism to 'Die Out,'" *BBC News,* October 24, 2003, http://newsvote.bbc.co.uk.
25. See "The Last Best Place," 60.
26. See Garrett Hardin, "The Tragedy of the Commons," *Science* 162 (1968): 1243–1248.
27. For an example of recent scholarship, see Oran R. Young, "Land Use, Environmental Change, and Sustainable Development: The Role of Institutional Diagnostics," *International Journal of the Commons* 5, no. 1 (February 2011): 66–85; for the key work in this field, see Elinor Ostrom, *Governing the Commons: The Evolution of Institutions for Collective Action* (Cambridge: Cambridge University Press, 1999).
28. UNDP, "Human Development Report: Mongolia 2003," 39.
29. Daniel J. Miller, "Tough Times for Tibetan Nomads in Western China: Snowstorms, Settling Down, Fences, and the Demise of

Traditional Nomadic Pastoralism," *Nomadic Peoples* 4, no. 1 (2000): 85.

30. Sneath, "Land Use," 449.
31. On the Mongolian term *zah zeel* for "the market," see the interesting discussion in Alan Wheeler, "Moralities of the Mongolian 'Market': A Genealogy of Trade Relations and the *Zah Zeel*," in *Inner Asia* 6, no. 2 (2004): 215–238.
32. Information is drawn from Anne Goes, "Index-Based Livestock Insurance in Mongolia," a paper presented at the April 2005 Mongolia Society Panel at the Association of Asian Studies annual meeting in Chicago. See also "Mongolia: Index-Based Livestock Insurance Project," document posted on the World Bank website, http://www-wds.worldbank.org (2005).
33. Miller, "Tough Times," 104.
34. See, for instance, "HRW Report Slams Tibetan Resettlements," *Tibet News Digest,* June 9–22, 2007, www.tibetinfonet.net; and "12,000 Pastoralist to Be Settled in Yunnan," *Tibet News Digest,* December 6–19, 2008, www.tibetinfonet.net.
35. See J. Marc Foggin, "Depopulating the Tibetan Grasslands," *Mountain Research and Development* 28, no. 1 (2008): 26–31.
36. Julia A. Klein, Emily Yeh, Joseph Bump, Yonten Nyima, and Kelly Hopping, "Coordinating Environmental Protection and Climate Change Adaptation Policy in Resource-Dependent Communities: A Case Study from the Tibetan Plateau," in *Climate Change Adaptation in Developed Nations: From Theory to Practice*, eds. James D. Ford and Lea Berrang-Ford (New York: Springer, 2011), 423–438.
37. Dee Mack Williams, *Beyond Great Walls: Environment, Identity, and Development on the Chinese Grasslands of Inner Mongolia* (Stanford, CA: Stanford University Press, 2002); David Sneath, "Land Use," 450–451; see also the review of Williams's book by Elizabeth Endicott in *China Review International* 10, no. 1 (Spring 2003): 272–277.
38. See Richard Fraser, "Forced Relocation amongst the Reindeer-Evenki of Inner Mongolia," *Inner Asia* 12, no. 2 (2010): 317–346.
39. Fraser, "Forced Relocation," 339.
40. Fraser, "Forced Relocation," 340.
41. D. Narantuya, "Giving the Land to the people," *MM* 19:617 (May 7, 2003), 1.
42. "President Discusses Land," *MM* (May 7, 2003), http://www.mongolnet.mn.
43. See "Law of Mongolia on Land," Article 8, http://www.investmongolia.com/law28.pdf.

44. See "Law of Mongolia on Land," Articles 20–21, specifically 21.4.4 and 21.5.6.
45. See "Law of Mongolia on Land," Article 54.2.
46. "Law of Mongolia on Land," Article 54.5.
47. Ministry of Food and Agriculture, "Food and Agriculture Policy of Government of Mongolia," (2005) http://gate1.pmis.gov.mn/mofa/english/idea_rural.htm.
48. "Law of Mongolia on Land," Article 48.1.
49. Mearns, "Decentralisation," 144.
50. "Law of Mongolia on Land," Article 6.2.
51. "Law of Mongolia on Land," Article 54.10.
52. "Law of Mongolia on Land," Article 54.2.
53. See Open Society Forum, *The Future of Nomadic Pastoralism in Mongolia: Public Perception Survey* (Ulaanbaatar: Open Society Forum, 2004), 40.
54. Ole Bruun, *Precious Steppe: Mongolian Nomadic Pastoralists in Pursuit of the Market,* (New York: Lexington Books, 2006), 211.
55. Bruun, *Precious Steppe*, 210.
56. On the limitations of sum governors' authority, see Bruun, *Precious Steppe*, 212; for the building of small cabins on pastures, see Bruun, *Precious Steppe*, viii, 52, 70, 76, and 213.
57. Natasha Fijn, *Living with Herds: Human-Animal Coexistence in Mongolia,* (New York: Cambridge University Press, 2011), 63.
58. Bruun, *Precious Steppe*, 213.
59. See Jörg Janzen, "Changing Political Regime and Mobile Livestock Keeping in Mongolia," *Geography Research Forum* 25 (2005), 67.
60. See Jörg Janzen, "Changing Political Regime and Mobile Livestock Keeping in Mongolia," *Geography Research Forum* 25 (2005): 68, for "camp communities"; see Robin Mearns, "Decentralisation," 140, for the more detailed definition.
61. See Janzen, "Changing Political Regime," 67 and 74.
62. Janzen, "Changing Political Regime," 77.
63. Janzen, "Changing Political Regime," 74.
64. Janzen, "Changing Political Regime," 70.
65. Open Society Forum, *Future of Nomadic Pastoralism*, 41.
66. Open Society Forum, *Future of Nomadic Pastoralism*, 126, Annex 1, Table 4.
67. See Tony Banks, "Property Rights and the Environment in Pastoral China: Evidence from the Field," *Development and Change* 32, no. 4 (2001): 717–740.
68. Banks, "Property Rights," 718.
69. Banks, "Property Rights," 773 and 736.

70. See Emily T. Yeh, "Tibetan Range Wars: Spatial Politics and Authority on the Grasslands of Amdo," *Development and Change* 34, no. 3 (2003): 512.
71. Yeh, "Tibetan Range Wars," 515, 516. Quotation from page 515.
72. Yeh, "Tibetan Range Wars," 516–517.
73. Yeh, "Tibetan Range Wars," 518.
74. See Shinjilt, "Pasture Fights, Mediation, and Ethnic Narrations: Aspects of the Ethnic Relationship between the Mongols and the Tibetans in Qinghai and Gansu," in *The Mongolia-Tibet Interface: Opening New Research Terrains in Inner Asia*, eds. Uradyn E. Bulag and Hildegard G. M. Diemberger (Boston, MA: Brill, 2007), 337–361.
75. See Bruun, *Precious Steppe*, 132–137.
76. Bruun, *Precious Steppe*, 235.
77. See Fijn, *Living with Herds*, 33, 222–223, 231–232.
78. Narantuya Danzan, *Religion in 20th Century Mongolia: Social Changes and Popular Practices* (Saarbrücken, Germany: VDM Verlag, 2008), 27.
79. Narantuya, *Religion*, 19.
80. John Perkins, *To the Ends of the Earth: Four Expeditions to the Arctic, the Congo, the Gobi, and Siberia* (New York: Pantheon Books, 1981), 92.
81. Walther Heissig, *Religions of Mongolia*, trans. Geoffrey Samuel (Berkeley, CA: University of California Press, 1980), 104.
82. Heissig, *Religions of Mongolia*, 43.
83. Vesna A. Wallace, "Mediating the Power of Dharma: The Mongols' Approaches to Reviving Buddhism in Mongolia," *The Silk Road* 6, no. 1 (Summer 2008): 46.
84. Narantuya, *Religion*, 105.
85. Wallace, "Mediating the Power of Dharma," 47.
86. Narantuya, *Religion*, 102–103.
87. Sant Maral Foundation, "Politbarometer #2" 35 (October 2008), in *MM* 50:908 (November 26, 2008), 9.
88. Sant Maral Foundation, "Politbarometer #8" 41 (October 2010).
89. Krisztina Teleki, "Building on Ruins, Memories and Persistence: Revival and Survival of Buddhism in the Mongolian Countryside," *The Silk Road* 7 (2009), 64.
90. See Zsuzsa Majer and Krisztina Teleki, "Buddhism in Mongolia Today," *The Silk Road* (online journal) 7 (2009), 51; and Teleki, "Building on Ruins," 70.
91. Zsuzsa Majer, "Continuation or Disjuncture with the Past and the Tibetan Buddhist Tradition," *The Silk Road* 7 (2009), 56.

92. Majer, "Continuation or Disjuncture," 57.
93. Majer, "Continuation or Disjuncture," 60.
94. Ts. Namkhainyambuu, *Bounty from the Sheep: Autobiography of a Herdsman,* trans. Mary Rossabi (Cambridge, UK: The White Horse Press, 2000), 89.
95. See Dennis Ojima and Togtohyn Chuluun, "Policy Changes in Mongolia: Implications for Land Use and Landscapes," in *Fragmentation in Semi-Arid and Arid Landscapes: Consequences for Human and Natural Systems*, ed. K. A. Galvin, R. S. Reid, R. H. Behnke, Jr., and N. T. Hobbs (Dordrecht, The Netherlands: Springer, 2010), 187–188.
96. See "Draft Law of Mongolia on Pasture Land," at www.greengold.mn/media/ . . . /091103%20Pasture%20Law%20Draft_eng.pdf. Article 20 relates to the settlement of pasture disputes.
97. See Open Society Forum, *Future of Nomadic Pastoralism*, 39.
98. Mari Kazato, "Local Value and Rights to Winter Camps under Land Privatization Policy in Postsocialist Mongolia," paper presented at the 11th Biennial Conference of IASCP (International Association for the Study of Common Property), in Bali, Indonesia (June 19–23, 2006), 3.
99. Fijn, *Living with Herds*, 183.
100. B. Shirendyb, "Some Aspects of the History of Land Rights in Mongolia," *Journal of the Anglo-Mongolian Society* 3, no. 2 (1976): 44.
101. Maria E. Fernandez-Gimenez, "Spatial and Social Boundaries and the Paradox of Pastoral Land Tenure: A Case Study from Postsocialist Mongolia," *Human Ecology: An Interdisciplinary Journal* 30, no. 1 (March 2002): 57.
102. Kazato, "Local Value and Rights to Winter Camps," 7.
103. This section makes use of Fernandez-Gimenez, "Spatial and Social Boundaries," 54, 56; and Kazato, "Local Value and Rights to Winter Camps," 6–7.
104. See David Sneath, "Property Regimes and Sociotechnical Systems: Rights over Land in Mongolia's 'Age of the Market,'" in *Property in Question: Value Transformation in the Global Economy*, eds. Katherine Verdery and Caroline Humphrey (New York: Berg, 2004), 164–165.
105. See Robin Mearns, "Decentralisation, Rural Livelihoods and Pasture-Management in Post-Socialist Mongolia," *European Journal of Development Research* 16, no. 1 (Spring 2004): 142–144.
106. Fernandez-Gimenez, "Spatial and Social Boundaries," 56.
107. Kazato, "Local Value and Rights to Winter Camps," 10.

108. Kazato, "Local Value and Rights to Winter Camps," 10.
109. Sant Maral Foundation, "Politbarometer #8" 41 (October 2010).
110. Sneath, "Property Regimes," 166.
111. B. Altantuul, "Ikh Khural Accelerates Pace, Approves Land Privatization," *MTW,* July 2, 2002.
112. See Sant Maral Foundation, "Politbarometer #17" (October 2002).
113. See Maria E. Fernandez-Gimenez and B. Batbuyan, "Law and Disorder: Local Implementation of Mongolia's Land Law, " *Development and Change* 35, no. 1 (January 2004): 154.
114. Fernandez-Gimenez and Batbuyan, "Law and Disorder," 156.
115. Mearns, "Decentralisation, Rural Livelihoods," 141.
116. See "Herders Want Ownership of Pastureland," *UBP,* December 16, 2004.
117. See Dee Mack Williams, *Beyond Great Walls.*
118. Mukhamet Shayakhmetov, *The Silent Steppe: The Memoir of a Kazakh Nomad under Stalin*, trans. from the Russian by Jan Butler (New York: The Rookery Press, 2007), 4.
119. Shayakhmetov, *Silent Steppe*, 5.
120. Shayakhmetov, *Silent Steppe*, 3.
121. Iliya I. Alimaev, Roy H. Behnke, Jr., "Ideology, Land Tenure and Livestock Mobility in Kazakhstan," in *Fragmentation in Semi-Arid and Arid Landscapes: Consequences for Human and Natural Systems,* ed. K. A. Galvin, R. S. Reid, R. H. Behnke, Jr. and N. T. Hobbs (Dordrecht, The Netherlands: Springer, 2010), 155–156.
122. See Alimaev and Behnke, "Ideology, Land Tenure and Livestock Mobility," 157; see also Martha Brill Olcott, *The Kazakhs* (Stanford, CA: Stanford University, Hoover Institution Press, 1987), 98.
123. Shayakhmetov, *Silent Steppe*, 6.
124. Alimaev and Behnke, "Ideology, Land Tenure and Livestock Mobility," 153.
125. The following draws upon Virginia Martin's findings in her *Law and Custom in the Steppe: The Kazakhs of the Middle Horde and Russian Colonialism in the Nineteenth Century* (Richmond, Surrey, UK: Curzon, 2001), Chapter Five.
126. Olcott, *Kazakhs*, 184–185.
127. Olcott, *Kazakhs*, 187.
128. See the interesting discussion of contributing factors in Alimaev and Behnke, "Ideology, Land Tenure, and Livestock Mobility," 163–165.
129. Sarah Robinson, Peter Finke, and Bettina Hamann, "The Impacts of De-Collectivization on Kazak Pastoralists. Case Studies from

Kazakstan, Mongolia and the People's Republic of China," *Journal of Central Asian Studies* 4, no. 2 (Spring/Summer 2000): 5–6.

130. Namkhainyambuu, *Bounty from the Sheep.*
131. Robinson et al., "Impacts of De-Collectivization," 7.
132. Alimaev and Behnke, "Ideology, Land Tenure, and Livestock Mobility," 169–170.
133. Robinson et al., "Impacts of De-Collectivization," 7–8.
134. David Sneath, "Land Use, the Environment and Development in Post-Socialist Mongolia," *Oxford Development Studies* 31, no. 4 (December 2003): 442.
135. See the summary of the conference discussions in *The Kazakhstan Livestock Sector in Transition to a Free Market*, eds. Anatoly Khazanov, Vitaly Naumkin, Kenneth Shapiro, and David Tomas (Moscow: Russian Center for Strategic Research and International Studies, 1999), 295–301. I am indebted to Professor Khazanov for giving me a copy of this volume.
136. Alimaev and Behnke, "Ideology, Land Tenure, and Livestock Mobility," 173.
137. Alimaev and Behnke, "Ideology, Land Tenure, and Livestock Mobility," 174.
138. See Nurlan Almaganbetov, "The Impact of Land Reform on Economic Development in Kazakhstan," Central European University Center for Policy Studies: Open Society Institute, 2004–2005.
139. See Y. Kawabata et al., "Uranium and Thorium Isotopes from Kazakhstan," *Journal of Radioanalytical and Nuclear Chemistry* 278, no. 2 (2008): 459–462.

6 Agriculture, Mining, Tourism: Competing Interests in Land Use

1. Over 50 years ago, Róna-Tas described the process in depth in an illustrated article that he wrote based on field research done in Mongolia in 1957. See A. Róna-Tas, "Some Data on the Agriculture of the Mongols," in *Opuscula Ethnologica Memoriae Ludovici Biró Sacra*, eds. Tibor Bodrogi and L. Boglar (Budapest: Akadémiai Kiadó, 1959), 443–472.
2. See photos # 18, 19, and 20 in Elizabeth Endicott, *Pages from the Past: The 1910 Moscow Trade Expedition to Mongolia* (Norwalk, CT: EastBridge, 2007), xxi-xxii. The Róna-Tas article also provides excellent photos and drawings of agricultural implements.

3. Endicott, *Pages from the Past*, 135–136.
4. Owen Lattimore, "The Geographical Factor in Mongol History," originally published in *The Geographical Journal* 91, no. 1 (January 1938); republished in Owen Lattimore, *Studies in Frontier History: Collected Papers 1921–58* (London and New York: Oxford University Press, 1962), 257.
5. Lattimore, "Geographical Factor," 258.
6. Lattimore, "Geographical Factor," 258.
7. David Sneath, "The Rural and the Urban in Pastoral Mongolia," in *Mongols: From Country to City*, eds. Ole Bruun and Li Narangoa (Copenhagen: Nordic Institute of Asian Studies, 2006), 151.
8. William T. Rowe, "Owen Lattimore, Asia, and Comparative History," *The Journal of Asian Studies* 66, no. 3 (August 2007): 764.
9. Lattimore, "Geographical Factor," 241.
10. Lattimore, "Geographical Factor," 245.
11. See Anatoly M. Khazanov, *Nomads and the Outside World*, 2nd ed. (Madison, WI: University of Wisconsin Press, 1994), Chapter 2.
12. Lattimore, "Geographical Factor," 247.
13. Owen Lattimore, "On the Wickedness of Being Nomads," originally published in *T'ien Hsia Monthly* (Shanghai) 1, no. 2 (September 1935); republished in Lattimore, *Studies in Frontier History*, 417.
14. Lattimore, "Geographical Factor," 241.
15. Lattimore, "Geographical Factor," 245.
16. "China Steps Up Resettlement of Tibetan Herders," *Tibet News Digest*, January 17, 2008, Info@TibetInfoNet.net.
17. See Daniel J. Miller, "Tough Times for Tibetan Nomads in Western China: Snowstorms, Settling Down, Fences and the Demise of Traditional Nomadic Pastoralism," *Nomadic Peoples* NS 4, no. 1 (2000): 83–109.
18. MunkhDalai A. Zhang, Elles Borjigin, and Huiping Zhang, "Mongolian Nomadic Culture and Ecological Culture: On the Ecological Reconstruction in the Agro-Pastoral Zone in Northern China," *Ecological Economics* 62 (2007): 23.
19. Zhang et al., "Mongolian Nomadic Culture," 20.
20. Zhang et al., "Mongolian Nomadic Culture," 21.
21. Zhang et al., "Mongolian Nomadic Culture," 24.
22. Lattimore, "On the Wickedness of Being Nomads," 420.
23. Owen Lattimore, *Nomads and Commissars: Mongolia Revisited* (New York: Oxford University Press, 1962), 26.
24. Lattimore, *Nomads and Commissars*, 165.
25. Lattimore, *Nomads and Commissars*, 30.

26. M. Dash, "Innovations in Animal Husbandry," *Mongolia* (Ulaanbaatar) 2, no. 5 (March 1972): 9–10.
27. See "Agriculture," *Mongolia* (Ulaanbaatar) 1, no. 46 (1979): 7.
28. See Sh. Gungaadorj, "Introducing Mongolia: The Selenge Aimak," *Mongolia* (Ulaanbaatar) 2, no. 71 (1983): 11.
29. Reginald and Ann Hibbert, *Letters from Mongolia* (London and New York: The Radcliffe Press, 2005), 235.
30. See, for instance, Robert B. Cassidy, Jr., "Mongolia: At Plan's End," *Asian Survey* 11, no. 1 (January 1971); 86–91. Cassidy (86–87) notes that crop failures in Mongolia in 1968–1969 led to Mongolia's inability to export grain to the USSR in 1969.
31. Hibbert, *Letters from Mongolia*, 121.
32. Hibbert, *Letters from Mongolia*, 242.
33. Statistics from T. Bayarsaihan and T. J. Coelli, "Productivity Growth in Pre-1990 Mongolian Agriculture: Spiraling Disaster or Emerging Success?" *Agricultural Economics* 28 (2003): 123.
34. See Bayarsaihan and Coelli, "Productivity Growth in Pre-1990 Mongolian Agriculture," 121–137.
35. See the fascinating interviews with Lookhuuz conducted by Yuki Konagaya and I. Lkhagvasuren, "Tsogt-Ochiriin Lookhuuz: Former Director of State Farms," in *Socialist Devotees and Dissenters: Three Twentieth-Century Mongolian Leaders*, ed., compiled, and trans. by Morris Rossabi (Osaka: National Museum of Ethnology, 2011), 84–97.
36. See World Bank, *Mongolia Quarterly* (January 28, 2008), at http://siteresources.worldbank.org/INTMONGOLIA/Resources/MNQtrlyFeb2008.pdf 3, 9. China has been the main foreign supplier of fruits and vegetables in the postsocialist period.
37. "Russia to Export 100,000 Tons of Wheat to Mongolia," *MWN*, April 15, 2008.
38. "Mongolia Facing Huge Increase, Possibly as High as 35%, in Cereal," *MWN*, February 19, 2008; see also World Bank, *Mongolia Quarterly* (January 28, 2008), 11.
39. P. Shinebayar, "Heat Wave Affects Crop Industry," *UBP*, August 2, 2010. See also *Annual Report 2009* of the Swiss Agency for Development and Cooperation in Mongolia (SDC), available on the SDC website, http://www.swiss-cooperation.admin.ch/mongolia/
40. D. Jargalsaikhan, "Bear Service," *UBP*, September 3, 2010. Jargalsaikhan's source on wheat and flour imports was Oyungerel, an inspector in the General Customs Office of Mongolia.

41. See "Development Context," *Onsite Insights* (SDC) 52 (October 2010), available on the SDC website, http://www.swiss-cooperation.admin.ch/mongolia/
42. "Information on Mongolian Agriculture," paper by the Farmers Knowledge Association (NGO) for the Ministry of Food and Agriculture for the Foreign Investors' Forum in Mongolia (June 2004) www.owc.org.mn/fermerdem/english/monagro/crop_2004.html.
43. See "Information on Mongolian Agriculture"; and Douglas A. Johnson, Dennis P. Sheehy, Daniel Miller, and Daalkhaijav Damiran, "Mongolian Rangelands in Transition," *Sécheresse* 17, nos. 1–2 (2006): 133–141, esp. 137–139.
44. See Jeremy J. Swift, "Case Study: Institutionalizing Pastoral Risk Management in Mongolia: Lessons Learned," (Rome, 2007), http://www.fao.org/NR/clim/docs/clim_070601_en.pdf. Swift's case study is based upon a FAO (Food and Agriculture Organization of the United Nations) study.
45. See Johnson et al., "Mongolian Rangelands," 137–139.
46. "News-in-Brief from the Montsame Agency," *MM* 44:1009 (November 5, 2010), 2. See also the update by B. Ooluun, "President of Laos National Assembly Visits Mongolia," *MM* 09:1078 (March 2, 2012), 1, 3.
47. According to an article by M. Tumurjav of the Mongolian National Agricultural University, the short-legged livestock may not be able to find natural fodder if snow cover is deeper than 25–30 cm. See Myatavyn Tumurjav, "Traditional Animal Husbandry Techniques Practiced by Mongolian Nomadic People," in *Mongolia Today: Science, Culture, Environment and Development*, eds. D. Badarch, R. A. Zilinskas, and P. J. Balint (London and New York: RoutledgeCurzon, 2003), 91 and 106.
48. The "2005 Progress Report" of the IFAD Rural Poverty Reduction Programme in Mongolia, www.rprmongolia.mn/pdf/2005_annual_report_en.pdf.
49. See, for instance, Ch. Sumiyabazar, "Aid to Herders in Severe Winter Storm Areas," *UBP*, February 2008.
50. Swift, "Case Study," 14, 28.
51. B. Oyundelger, "Land Use Programme Comes under Scrutiny," *MM* 10:868 (March 5, 2008), 1.
52. "Virgin Land III Campaign," *MM* 16:874 (April 16, 2008), 3.
53. See B. Oyundelger, "2010 Harvest Is Underway!" *MM* 36:1001 (September 10, 2010), 7.

54. See "Economic and Social Situation of Mongolia," *MM* 3:915 (January 21, 2009), 6.
55. "Crops Harvested in "Third Campaign for Reclaiming Virgin Land," *MM* 52:910 (December 10, 2008), 4.
56. "ATAR III Improves Harvest Yield," *MM* 42:1007 October 22, 2010), 3.
57. "Better Ag Stems from ATAR-3 Success," *MM* 52:1017 (December 24, 2010), 9.
58. "Best Harvest in 20 Years," *MM* 37:1054 (September 9, 2011), 1.
59. "Better Ag Stems from ATAR-3 Success."
60. Richard P. Reading, Don Bedunah, and Sukh Amgalanbaatar, "Conserving Mongolia's Grasslands, with Challenges, Opportunities, and Lessons For North America's Great Plains," *Great Plains Research* 20 (Spring 2010): 92.
61. I am indebted to Dr. B. Erdenebolor of the Mongolian State University of Agriculture for sharing with me his presentations and publications from which this section draws its information unless noted otherwise: B. Erdenebolor, PowerPoint Presentation, "Is Semi-Intensive Livestock Farming an Alternative to Pastoralism?"; B. Erdenebolor, "Report of the Study 'Analysis of the Business Environment of Milk Production in the Ulaanbaatar Area and Identification of Necessary Contents of Advisory Services and Training for Dairy Farmers"; and Baast Erdenebolor and Volker Hoffmann, "Feasibility and Implementation Strategies of Dairy Extension in Ulaanbaatar/Mongolia," *Studies on the Agricultural and Food Sector in Central and Eastern Europe*, 52, Halle, (Saale), IAMO: 91–98. Dr. Erdenebolor and his wife, Dr. B. Nyamgerel, also a professor at the Mongolian State University of Agriculture, generously spent a day taking my husband and me on a tour of peri-urban dairy farms a few hours from downtown Ulaanbaatar in August 2011.
62. Law of Mongolia on Land/Revised Version/June 7, 2002, www.nvcmongolia.com;nvce/download/Land.pdf
63. See Millennium Challenge Account—Mongolia, http://www.mca.mn.
64. For a discussion of this problem, see Swift, "Case Study," 9–10, 37.
65. Correspondence from Sas Carey, July 15, 2010; conversation with Sas Carey August 27, 2010.
66. See Ch. Sumiyabazar, "Fresh Clashes in Uvs Province," *UBP*, July 30, 2010.

67. "Locals Attack Chinese Gold Mine," *MM* 32:997 (August 13, 2010), 2.
68. "Gold Extraction in Orkhon Basin Suspended," *MWN,* August 20, 2010.
69. Ch. Sumiyabazar, "Fresh Shootout at Gold Mines in Tunkhel Village," *UBP,* September 7, 2010.
70. See Stefan Lovgren, "Mongolia Gold Rush Destroying Rivers, Nomadic Lives," *National Geographic News,* October 17, 2008, http://news.nationalgeographic.com/news/pf/10277748.html; and Sh. Batmonkh, "Protesters Threaten Violence to Save River," *UBP,* May 29, 2008.
71. "One Dies as Herders and Ninjas Clash in Gobi-Altai," *MWN,* August 20, 2010.
72. See "SDC Hands over Processing Plant Foundations to Soum Government," on the SDC Mongolia website, http://www.swiss-cooperation.admin.ch/mongolia/ed/Home/News/. This is just one example of the SDC intervention sites that aid artisanal miners and develop rehabilitation plans for previously mined lands. See also the *Annual Report 2009* of the SDC in Mongolia, available on the SDC website, http://www.swiss-cooperation.admin.ch/mongolia/
73. Louisa Lim, "Mongolians Seek Fortune in Gold, but at a Cost," *NPR online,* September 7, 2009, http://www.npr.org.
74. B. Ooluun, "Coal Leads Mongolia's Mining Industry," *MM* 07–08:1076–1077 (February 17, 2012), 1; "Mongolia Leads Coking Coal Exports to China in 2011," *MM* 07–08:1076–1077 (February 17, 2012), 12.
75. See Reading et al., "Conserving Mongolia's Grasslands," 95.
76. P. Shinebayar, "Every Citizen of Mongolia Would Own 550 Shares in Tavantolgoi," *UBP,* January 21, 2011; B. Ooluun, "One Preferential Share of Erdenes MGL for Each Citizen," *MM* 15:1032 (April 15, 2011), 6.
77. "Herders Take to the Streets, Four Arrested," Southern Mongolian Human Rights Information Center website, http//:smhric.org/news (May 26, 2011). This website provided photos of police arrests and beatings of demonstrators. See also Jonathan Watts, "Herder's Death Deepens Tensions in Inner Mongolia," *The Guardian,* May 27, 2011, http://www.guardian.co.uk/world/2011/may/27/tensions-miners-inner-mongolia. On the trial and sentencing of the coal truck driver, see "Death Sentence over Inner Mongolia Herder Killing," BBC news, June 8, 2011, http://www.bbc.co.uk/news/world-asia-pacific-13697612.

78. See Andrew Russell and Gillian Wallace, "Irresponsible Ecotourism," *Anthropology Today* 20, no. 3 (June 2004): 2.
79. Both quotations from "Tourism Director Wants More Well-Informed Guides," Official Tourism Website of Mongolia, http://www.mongoliatourism.gov.mn (August 27, 2010).
80. See BirdLife Asia, "Safeguarding Important Areas of Natural Habitat in Mongolia alongside Economic Development," *World Bank: Mongolia Discussion Papers, East Asia and Pacific Region Sustainable Development Department*, Washington, DC (January 2009), http://siteresources.worldbank.org/INTMONGOLIA/Resources/1_Safeguarding_important_areas_ENG.pdf 37.
81. Buyandelger Ulziikhuu, *Rural Development in Mongolia; Role of Tourism in Local Prosperity: A Case of Uvurkhangai Province* (Saarbrücken, Germany: LAP LAMBERT, 2010), 18.
82. Information is drawn from the website of the Gün-Galuut Nature Reserve, http://www.argalipark.com; and from the website of the Steppe Nomads Tourist Camp, http://www.mongoliagercamp.com.
83. See "Official Tourism Website of Mongolia," http://www.mongoliatourism.gov.mn.
84. Ulziikhuu, *Rural Development*, 58–59.
85. The following information is drawn from Nomadic Expeditions' website, http://www.nomadicexpeditions.com; from my two visits (2006, 2010) to the Three Camel Lodge; and from "Mongol Nutag: Mongolia Deep in My Soul," (interview with Jalsa Urubshurow, founder of Nomadic Expeditions) in *Smart Way: The Inflight Magazine of Eznis Airways* 2, no. 012 (2011): 38–43.
86. BirdLife Asia, "Safeguarding Important Areas," 73.
87. See the website of "Pack for a Purpose," http://www.packforapurpose.org.
88. See the website of "Ikh Nart Nature Reserve," http://www.ikhnart.com/ecotourism.html; and Reading et al., "Conserving Mongolia's Grasslands," 98.
89. Ulziikhuu, *Rural Development*, 60.
90. See Morris Rossabi, *Modern Mongolia: From Khans to Commissars to Capitalists* (Berkeley, CA: University of California Press, 2005), 129, for criticisms of foreign aid donors' work in the herding sector.
91. Information on the takhi, the Khustai National Park, and the ger camp is drawn from the website of the Foundation for the Preservation and Protection of the Przewalski Horse, http:/www.treemail.nl/takh/index.htm
92. Reading et al., "Conserving Mongolia's Grasslands," 98.

93. See "Support for community based tourism development in Eej Khairkhan protected area," on the GEF Small Grants Programme website, http://sgp.undp.org/web/projects.
94. Information from "Herders Contribute to Conservation in the Altai Mountains in Mongolia," *UNDP Mongolia: News: Sustainable Natural Resource Management*, http://www.undp.mn/news-snrm-200510.html.
95. These and other Mongolia projects are listed on the GEF Small Grants Programme website, http://sgp.undp.org/web/projects.
96. See the SDC website, http://www.swiss-cooperation.admin.ch/mongolia/en/Home/Development_Cooperation/Mongolian_Potato_Programme.
97. See the description of the "Gobi Forage" Project on the Mercy Corps webpage, http://www.mercycorps.org.mn.
98. Information on the Green Gold Project is drawn from the *Annual Report 2009* of the SDC in Mongolia, available from the SDC website. The Draft Law on Pastureland will be discussed in the concluding chapter.
99. The SDC, *Annual Report 2009*, 7. For further information on the PUGs and the current legal climate in Mongolia, see also 24.

7 Concluding Thoughts

1. See P. Shinebayar, "Livestock Number Down," *UBP,* January 6, 2011. Shinebayar cites the National Statistical Office of Mongolia.
2. These and other statistics that follow are from the Swiss Agency for Development and Cooperation (SDC) *Annual Report 2009*, 8.
3. SDC, *Onsite Insights* 63 (October 2011), available on the SDC website, http://www.swiss-cooperation.admin.ch/mongolia/
4. Morris Rossabi, *Modern Mongolia: From Khans to Commissars to Capitalists,* (Berkeley, CA: University of California Press, 2005), 129.
5. See Shafqat Hussain, "Protecting the Snow Leopard and Enhancing Farmers' Livelihoods: A Pilot Insurance Scheme in Baltistan," *Mountain Research and Development* 20, no. 3 (August 2000): 226–231.
6. *Changing Nomads in a Changing World*, ed. Joseph Ginat and Anatoly M. Khazanov (Portland, Oregon: Sussex Academic Press, 1998), 12.
7. See Reginald and Ann Hibbert, *Letters from Mongolia* (London and New York: The Radcliffe Press, 2005).

8. D. Jargalsaikhan, "Democracy and Dictatorship," *UBP*, August 20, 2010.
9. Jargalsaikhan, "Democracy and Dictatorship."
10. "Nomads No More: A Steppe-Land Struggles with New Riches," *The Economist* (October 21, 2010), http://www.economist.com/node/17312256?story_id=17312256.
11. "Nomads No More."
12. See "Jargal de Facto: More Grants, Less Labour in the Mongolian Desert," *Mongolia Web News*, October 11, 2011, http://www.mongolia-web.com.
13. See Troy Sternberg, "Unravelling Mongolia's Extreme Winter Disaster of 2010," *Nomadic Peoples* 14, no. 1 (2010): 83–84.
14. Gregory Delaplace, "Chinese Ghosts in Mongolia," *Inner Asia* 12, no. 1 (2010): 127–141.
15. Delaplace, "Chinese Ghosts," 135.
16. Jeffrey Reeves, "Mongolia's Environmental Security: Chinese Unconscious Power and Ulaanbaatar's State Weakness," *Asian Survey* 51, no. 3 (May/June, 2011): 454.
17. B. Oyundelger, "Mongolia's Coal to Reach World Market," *MM* 44:1009 (November 5, 2010), 9.
18. See "Special Report: Mongolia's Fabled Mine Stirs Asian Frontier," Reuters (October 12, 2010), http://www.alternet.org/thenews/newsdesk/SGE685S060.htm
19. Sant Maral Foundation, "Politbarometer #8" 41 (October 2010). The Politbarometer surveys are available at http://www.santmaral.mn/en/publications.
20. Sant Maral Foundation, "Politbarometer #9" 42 (May, 2011).
21. Robin Mearns, "Decentralisation, Rural Livelihoods and Pasture-Land Management in Post-Socialist Mongolia," *European Journal of Development Research* 16, no. 1 (Spring 2004): 141; and Ole Bruun, *Precious Steppe: Mongolian Nomadic Pastoralists in Pursuit of the Market* (New York: Lexington Books, 2006), 237.
22. See Bruun, *Precious Steppe*, 52, 70, and 213 on the building of such small cabins in Arkhangai Aimag.
23. "Draft Law of Mongolia on Pasture Land," www.greengold.mn/media/...091103%20Pasture%20Law%20Draft_eng.pdf.
24. Dorligsuren et al., "Lessons from a Territory-Based Community Development Approach in Mongolia: Ikhtamir Pasture User Groups," in *Restoring Community Connections to the Land: Building Resilience through Community-Based Rangeland Management in China and Mongolia*, eds. Maria E. Fernandez-Gimenez, Xiaoyi

Wang, B. Batkhishig, Julia A. Klein, and Robin S. Reid (Cambridge, MA: CAB International, 2012), 184–185.

25. Information from Millennium Challenge Account Mongolia website, http://www.mca.mn/?=eng/Project/PeriUrban.
26. See, for instance, Uradyn E. Bulag, "Mongolia in 2008: From Mongolia to Mine-golia," *Asian Survey* 49, no. 1 (Jan/Feb 2009): 129–134.
27. "Foreign Firms Aim to Win Big on Mongolia Coal Mine," *UBP*, April 15, 2011.
28. For the concept of Mongolia as "land-linked" and for a discussion of the "new Great Game," see Uradyn E. Bulag, "Mongolia in 2009: From Landlocked to Land-linked Cosmopolitan," *Asian Survey* 50, no. 1 (Jan/Feb 2010): 97–103.
29. Thomas Barfield, "Nomadic Pastoralism in Mongolia and Beyond," in *Mapping Mongolia: Situating Mongolia in the World from Geologic Time to the Present*, ed. Paula L. W. Sabloff (Philadelphia, PA: University of Pennsylvania Museum of Archaeology and Anthropology, 2011), 122.

Bibliography

Books

Academy of Sciences of the MPR. *Information Mongolia*. New York: Pergamon Press, 1990.

Allsen, Thomas T. *The Royal Hunt in Eurasian History*. Philadelphia, PA: University of Pennsylvania Press, 2006.

Andrews, Roy Chapman. *Across Mongolian Plains: A Naturalist's Account of China's "Great Northwest."* New York: Blue Ribbon Books, 1921.

Atwood, Christopher P. *Encyclopedia of Mongolia and the Mongol Empire*. New York: Facts on File, 2004.

Avery, Martha. *The Tea Road: China and Russia Meet across the Steppe*. Beijing: China Intercontinental Press, 2003.

———. *Women of Mongolia*. Seattle, WA: University of Washington Press, 1996.

Bawden, C. R. *The Modern History of Mongolia*. New York: Praeger, 1968.

Bell, John. *A Journey from St. Petersburg to Pekin 1719–22*. New York: Barnes & Noble, 1966, reprint of the 1763 original.

Bold, Bat-Ochir. *Mongolian Nomadic Society: A Reconstruction of the "Medieval" History of Mongolia*. New York: St. Martin's Press, 2001.

Bruun, Ole. *Precious Steppe: Mongolian Nomadic Pastoralists in Pursuit of the Market*. New York: Lexington Books, 2006.

Danzan, Narantuya. *Religion in 20th Century Mongolia: Social Changes and Popular Practices*. Saarbrücken, Germany: VDM Verlag, 2008.

Endicott, Elizabeth. *Pages from the Past: The 1910 Moscow Trade Expedition to Mongolia*. Norwalk, CT: EastBridge, 2007.

Endicott-West, Elizabeth. *Mongolian Rule in China: Local Administration in the Yuan Dynasty*. Cambridge: Harvard University Press, 1989.

Erdeni-yin tobči: Geschichte der Ost-Mongolen und ihres Fürstenhauses verfasst von Ssanang Ssetsen Chungtaidschi der Ordus. I. J. Schmidt edition. Beijing, 1937, reprint of the St. Petersburg, N Gretsch, 1829 ed.

Fijn, Natasha. *Living with Herds: Human-Animal Coexistence in Mongolia*. New York: Cambridge University Press, 2011.

Finch, C. *Mongolia's Wild Heritage: Biological Diversity, Protected Areas, and Conservation in the Land of Chingis Khaan*. Ulaanbaatar: Mongolia Ministry for Nature and Environment, 1996.

Gardelle, Linda. *Pasteurs nomades de Mongolie: Des sociétés nomades et des États*. Paris: Buchet/Chastel, Écologie, 2010.

Goldstein, Melvyn C. and Cynthia M. Beall. *The Changing World of Mongolia's Nomads*. Berkeley, CA: University of California Press, 1994.

Graivoronskii, V. V. *Kooperirovannoe aratstvo MNR : izmeneniia v urovne zhizni 1960–1980*. Moscow: "Nauka," 1982.

———. Ot kochevogo obraza zhizni k osedlosti (na opyte MNR). Moscow: "Nauka," 1979.

———. *Reformy v sotsial'noi sfere sovremennoi Mongolii*. Moscow: Institut vostokovedeniia RAN, 2007.

Hansen, Valerie. *Negotiating Daily Life in Traditional China: How Ordinary People Used Contracts 600–1400*. New Haven, CT: Yale University Press, 1995.

Heissig, Walther. *A Lost Civilization: The Mongols Rediscovered*. Translated by D. J. S. Thomson. New York: Basic Books, 1966.

———. *The Religions of Mongolia*. Translated by Geoffrey Samuel. Berkeley, CA: University of California Press, 1980.

Hibbert, Reginald and Ann. *Letters from Mongolia*. London and New York: The Radcliffe Press, 2005.

Jagchid, Sechin and Paul Hyer. *Mongolia's Culture and Society*. Boulder, CO: Westview Press, 1979.

Kara, György. *Books of the Mongolian Nomads: More than Eight Centuries of Writing Mongolian*. Translated by John R. Krueger. Bloomington, IN: Indiana University Research Institute for Inner Asian Studies, 2005.

Khazanov, Anatoly M. et al., eds. *The Kazakhstan Livestock Sector in Transition to a Free Market*. Moscow: Russian Center for Strategic Research, 1999.

———. *Nomads and the Outside World*. 2nd ed. Madison, WI: University of Wisconsin Press, 1994.

Lattimore, Owen. *Nomads and Commissars: Mongolia Revisited*. New York: Oxford University Press, 1962.

Liu, Xiaoyuan. *Reins of Liberation: An Entangled History of Mongolian Independence, Chinese Territoriality, and Great Power Hegemony, 1911–1950*. Stanford, CA: Stanford University Press, 2006.

Marsh, Christopher. *Religion and the State in Russia and China: Suppression, Survival, and Revival*. London: Continuum International Publishing Group, 2011.

Martin, Virginia. *Law and Custom in the Steppe: The Kazakhs of the Middle Horde and Russian Colonialism in the Nineteenth Century*. Richmond, Surrey, UK: Curzon, 2001.

May, Timothy. *The Mongol Art of War*. Yardley, PA: Westholme Publishing, 2007.

Moskovskaia torgovaia ekspeditsiia v Mongoliiu. Moscow: P. P. Riabushinskii Printing House, 1912.

Namkhainyambuu, Ts. *Bounty from the Sheep: Autobiography of a Herdsman*. Translated by Mary Rossabi. Cambridge, UK: The White Horse Press, 2000.

Olcott, Martha Brill. *The Kazakhs*. Stanford, CA: Stanford University Press, 1987.

Open Society Forum. *The Future of Nomadic Pastoralism in Mongolia: Public Perception Survey*. Ulaanbaatar: Open Society Forum, 2004.

Ostrom, Elinor. *Governing the Commons: The Evolution of Institutions for Collective Action*. Cambridge: Cambridge University Press, 1999.

Paine, S. C. M. *Imperial Rivals: China, Russia, and Their Disputed Frontier*. Armonk, NY: M. E. Sharpe, 1996.

Pedersen, Morten Axel. *Not Quite Shamans: Spirit Worlds and Political Lives in Northern Mongolia*. Ithaca, NY: Cornell University Press, 2011.

Perdue, Peter C. *China Marches West: The Qing Conquest of Central Eurasia*. Cambridge: Harvard University Press, 2005.

Perkins, John. *To the Ends of the Earth*. New York: Pantheon Books, 1981.

Perry-Ayscough, H. G. C. and Captain R. B. Otter-Barry. *With the Russians in Mongolia*. New York: John Lane Co., 1914.

Rasidondug, Š. and Veronika Veit. Petitions of Grievances Submitted by the People (18th – beginning of 20th century). Wiesbaden: Harrassowitz, 1975.

Rossabi, Morris. *Modern Mongolia: From Khans to Commissars to Capitalists*. Berkeley, CA: University of California Press, 2005.

Rupen, Robert. *How Mongolia Is Really Ruled: A Political History of the Mongolian People's Republic 1900–1978*. Stanford, CA: Stanford University Press, 1979.

Sambuu, Jamsrangiin. *Herdsman to Statesman. The Autobiography of Jamsrangiin Sambuu of Mongolia*. Translated from the Mongolian by Mary Rossabi. New York: Rowman & Littlefield, 2010.

Sanjdorj, M. *Manchu Chinese Colonial Rule in Northern Mongolia*. Translated from the Mongolian and annotated by Urgunge Onon. New York: St. Martin's Press, 1980.

Shayakhmetov, Mukhamet. *The Silent Steppe: The Memoir of a Kazakh Nomad under Stalin*. Translated by Jan Butler. New York: The Rookery Press, 2007.

Simukov, A. D. *Trudy o Mongolii i dlia Mongolii*. 3 vols. Edited by Yuki Konagaya, Sanzhaasurengiin Bayarlaa, and Ichinkhorloogiin Lkhagvasuren. Osaka: The National Museum of Ethnology, 2007–2008.

Steinhauer-Burkart, Bernd. *Gobi Gurvansaikhan National Park*. Ulaanbaatar: Eco-Nature, 1999.

The Secret History of the Mongols: A Mongolian Epic Chronicle of the Thirteenth Century. Translated with a historical and philological commentary by Igor de Rachewiltz. 2 vols. Leiden and Boston: Brill, 2004.

Ulziikhuu, Buyandelger. *Rural Development in Mongolia: Role of Tourism in Local Prosperity: A Case of Uvurkhangai Province*. Saarbrücken, Germany: LAP LAMBERT, 2010.

Williams, Dee Mack. *Beyond Great Walls: Environment, Identity, and Development on the Chinese Grasslands of Inner Mongolia*. Stanford, CA: Stanford University Press, 2002.

Journal Articles and Book Chapters

Alimaev, Iliya I. and Roy H. Behnke, Jr. "Ideology, Land Tenure and Livestock Mobility in Kazakhstan." In *Fragmentation in Semi-Arid and Arid Landscapes: Consequences for Human and Natural Systems*. Edited by K. A. Galvin, R. S. Reid, R. H. Behnke, Jr. and N. T. Hobbs, 151–178. Dordrecht, The Netherlands: Springer, 2010.

Banks, Tony. "Property Rights and the Environment in Pastoral China: Evidence from the Field." *Development and Change* 32, no. 4 (2001): 717–740.

Barfield, Thomas. "Nomadic Pastoralism in Mongolia and Beyond." In *Mapping Mongolia: Situating Mongolia in the World from Geologic Time to the Present*. Edited by Paula L. W. Sabloff, 104–124. Philadelphia, PA: University of Pennsylvania Press, 2011.

Barkmann, Udo B. "Some Comments on the Consequences of the Decline of the Mongol Empire on the Social Development of the Mongols." In *The Mongol Empire and Its Legacy*. Edited by Reuven Amitai-Preiss and David O. Morgan, 273–281. Leiden: Brill, 1999.

Batbayar, Tsedendamba. "Mongolian-Russian Relations in the Past Decade." *Asian Survey* 43, no. 6 (Nov/Dec 2003): 951–970.

Bauer, Kenneth M. "Common Property and Power: Insights from a Spatial Analysis of Historical and Contemporary Pasture Boundaries among Pastoralists in Central Tibet." *Journal of Political Ecology* 13 (2006): 24–47.

Bawden, C. R. "A Joint Petition of Grievances Submitted to the Ministry of Justice of Autonomous Mongolia in 1919." *Bulletin of the School of Oriental and African Studies, University of London* 30, no. 3 (1967): 548–563.

Bayarsaikhan, T. and T. J. Coelli. "Productivity Growth in Pre-1990 Mongolian Agriculture: Spiraling Disaster or Emerging Success?" *Agricultural Economics* 28, no. 2 (2003): 121–137.

Bulag, Uradyn E. "Mongolia in 2008: From Mongolia to Mine-golia." *Asian Survey* 49, no. 1 (Jan/Feb 2009): 129–134.

———. "Mongolia in 2009: From Landlocked to Land-linked Cosmopolitan." *Asian Survey* 50, no. 1 (Jan/Feb 2010): 97–103.

Bumaa, Nasan Dashdendeviin. "The Twentieth Century: From Domination to Democracy." In *Modern Mongolia: Reclaiming Genghis Khan*. Edited by Paula L. W. Sabloff, 31–63. Philadelphia, PA: University of Pennsylvania Museum of Archaeology and Anthropology, 2001.

Cassidy, Robert B., Jr. "Mongolia: At Plan's End." *Asian Survey* 11, no. 1 (Jan 1971): 86–91.

Cleaves, F. W. "A Mongolian Rescript of the Fifth Year of Degedü Erdem-tü (1640)." *Harvard Journal of Asiatic Studies* 46, no. 1 (Jun 1986): 181–200.

Delaplace, Gregory. "Chinese Ghosts in Mongolia." *Inner Asia* 12, no.1 (2010): 127–141.

Dorligsuren, Dulamsuren, Batjav Batbuyan, Bulgamaa Densambu, and Steven R. Fassnacht. "Lessons from a Territory-Based Community Development Approach in Mongolia: Ikhtamir Pasture User Groups." In *Restoring Community Connections to the Land: Building Resilience through Community-Based Rangeland Management in China and Mongolia*. Edited by Maria E. Fernandez-Gimenez, Xiaoyi Wang, B. Batkhishig, Julia A. Klein, and Robin S. Reid, 166–188. Cambridge, MA: CAB International, 2012.

Endicott, Elizabeth. "The Mongols and China: Cultural Contacts and the Changing Nature of Pastoral Nomadism (Twelfth to Early Twentieth Centuries)." In *Mongols, Turks, and Others: Eurasian Nomads and the Sedentary World*. Edited by Reuven Amitai and Michal Biran, 461–481. Leiden: Brill, 2005.

———. "Beyond Great Walls: Environment, Identity, and Development on the Chinese Grasslands of Inner Mongolia (review)." *China Review International* 10, no. 1 (Spring 2003): 272–277.

Endicott-West, Elizabeth. "The Yüan Government and Society." In *Alien Regimes and Border States, 907–1368*, vol. 6 of *The Cambridge History of China*. Edited by Herbert Franke and Denis Twitchett, 587–615. New York: Cambridge University Press, 1994.

Erdenebolor, B. "Is Semi-Intensive Livestock Farming an Alternative to Pastoralism?" PowerPoint Presentation, 2011.

———. "Report of the Study 'Analysis of the Business Environment of Milk Production in the Ulaanbaatar Area and Identification of Necessary Contents of Advisory Services and Training for Dairy Farmers." Unpublished paper 2011.

Erdenebolor, B. and Volker Hoffmann. "Feasibility and Implementation Strategies of Dairy Extension in Ulaanbaatar/Mongolia." In *Multi-Level Processes of Integration and Disintegration: Proceedings of the Third Green Week Scientific Conference.* Edited by Franziska Schaft and Alfons Balmann. In *Studies on the Agriculture and Food Sector in Central and Eastern Europe* 52: 91–98.

Fernandez-Gimenez, Maria E. "Spatial and Social Boundaries and the Paradox of Pastoral Land Tenure: A Case Study from Postsocialist Mongolia." *Human Ecology: An Interdisciplinary Journal* 30, no. 1 (Mar 2002): 49–78.

Fernandez-Gimenez, Maria E. and B. Batbuyan. "Law and Disorder: Local Implementation of Mongolia's Land Law." *Development and Change* 35, no. 1 (Jan 2004): 141–166.

Foggin, J. Marc. "Depopulating the Tibetan Grasslands." *Mountain Research and Development* 28, no. 1 (Feb 2008): 26–31.

Fraser, Richard. "Forced Relocation amongst the Reindeer-Evenki of Inner Mongolia," *Inner Asia* 12, no. 2 (2010): 317–346.

Ginat, Joseph. "Preservation and Change in Bedouin Societies in Israel." In *Changing Nomads in a Changing World.* Edited by Joseph Ginat and Anatoly M. Khazanov, 171–201. Portland, OR: Sussex Academic Press, 1998.

Goes, Anne. "Index-Based Livestock Insurance in Mongolia." Unpublished paper presented at the April 2005 Mongolia Society Panel at the Association of Asian Studies annual meeting in Chicago.

Hardin, Garrett. "The Tragedy of the Commons." *Science* 162 (1968): 1243–1248.

Havstad, Kris, Jeffrey Herrick, and Enkh-Amagalan Tseelei. "Mongolia's Rangelands: Is Livestock Production the Key to the Future?" *Frontiers in Ecology and the Environment* 6, no. 7 (Sep 2008): 386–391.

Hussain, Shafqat. "Protecting the Snow Leopard and Enhancing Farmers' Livelihoods: A Pilot Insurance Scheme in Baltistan." *Mountain Research and Development* 20, no. 3 (Aug 2000): 226–231.

Janzen, Jörg. "Changing Political Regime and Mobile Livestock Keeping in Mongolia." *Geography Research Forum* 25 (2005): 62–82.

Johnson, Douglas A., Dennis P. Sheehy, Daniel Miller, and Daalkhaijav Damiran. "Mongolian Rangelands in Transition." *Sécheresse* 17, nos. 1–2 (2006): 133–141.

Johnstad, Mark D. and Richard P. Reading. "Mongolia's Protected Areas System." In *Biodiversity Briefings from Northern Eurasia* 2, no.1 (2003): 8–11.

Kawabata, Y. et al. "Uranium and Thorium Isotopes from Kazakhstan." *Journal of Radioanalytical and Nuclear Chemistry* 278, no.2 (2008): 459–462.

Kazato, Mari. "Local Value and Rights to Winter Camps Under Land Privatization Policy in Postsocialist Mongolia." Unpublished paper presented at the 11th Biennial Conference of IASCP (International Association for the Study of Common Property), Bali, Indonesia (Jun 19–23, 2006).

Khazanov, Anatoly M. "Pastoralists in the Contemporary World: The Problem of Survival." In *Changing Nomads in a Changing World*. Edited by Joseph Ginat and Anatoly M. Khazanov, 7–23. Portland, Oregon: Sussex Academic Press, 1998.

Klein, Julia A., Emily Yeh, Joseph Bump, Yonten Nyima, and Kelly Hopping. "Coordinating Environmental Protection and Climate Change Adaptation Policy in Resource-Dependent Communities: A Case Study from the Tibetan Plateau." In *Climate Change Adaptation in Developed Nations: From Theory to Practice*. Edited by James D. Ford and Lea Berrang-Ford, 423–438. New York: Springer, 2011.

Konagaya, Yuki and I. Lkhagvasuren. "Tsogt-Ochiriin Lookhuuz: Former Director of State Farms." In *Socialist Devotees and Dissenters: Three Twentieth-Century Mongolian Leaders*. Interviews conducted by Yuki Konagaya and I. Lkhagvasuren. Edited and compiled by Morris Rossabi. Translated by Mary Rossabi, 35–162. Osaka: National Museum of Ethnology, 2011.

Krätli, Saverio. "Editorial." *Nomadic Peoples* 14, no.1 (2010): 1–5.

Lan, Mei-hua. "China's 'New Administration' in Mongolia." In *Mongolia in the Twentieth Century: Landlocked Cosmopolitan*. Edited by Stephen Kotkin and Bruce A. Elleman, 39–58. Armonk, NY: M. E. Sharpe, 1999.

Lattimore, Owen. "The Geographical Factor in Mongol History." Originally published in *The Geographical Journal* XCI, no. 1 (Jan 1938); reprinted in Owen Lattimore. *Studies in Frontier History: Collected Papers 1921–58*, 241–258. London and New York: Oxford University Press, 1962.

———. "On the Wickedness of Being Nomads." Originally published in *T'ien Hsia Monthly* (Shanghai) I, no. 2 (Sep 1935); reprinted in Owen Lattimore. *Studies in Frontier History: Collected Papers 1921–58*, 415–426. London and New York: Oxford University Press, 1962.

Lazar, David. "Well Water Pumps Provide New Reliability and Water Supply for Mongolian Sheepherders." *Water Conditioning & Purification Magazine* (Sep 2004): 8–9.

Marx, Emanuel. "Are There Pastoral Nomads in the Arab Middle East?" In *The Anthropology of Tribal and Peasant Pastoral Societies: The Dialectics of Social Cohesion and Fragmentation*. Edited by Ugo Fabietti and Philip Carl Salzman, 101–115. Pavia: Collegio Ghislieri, 1996.

Mearns, Robin. "Decentralisation, Rural Livelihoods and Pasture-Land Management in Post-Socialist Mongolia." *European Journal of Development Research* 16, no. 1 (Spring 2004): 133–152.

Miller, Daniel J. "Tough Times for Tibetan Nomads in Western China: Snowstorms, Settling Down, Fences, and the Demise of Traditional Pastoral Nomadism." *Nomadic Peoples* 4, no. 1 (2000): 83–109.

Miller, Robert J. "Areas and Institutions in Eastern Asia." *Southwestern Journal of Anthropology* 9, no. 2 (Summer 1953): 203–211.

———. "Buddhist Monastic Economy: The Jisa Mechanism." *Comparative Studies in Society and History* 3, no. 4 (Jul 1961): 427–438.

"Mongol Nutag: Mongolia Deep in My Soul," (interview with Jalsa Urubshurow). In *Smart Way: The Inflight Magazine of Eznis Airways* 2, no. 012 (2011): 38–43.

Murphy, George. "Comment." *Comparative Studies in Society and History* 3, no. 4 (Jul 1961): 439–442.

Natsagdorj, Sh. "The Economic Basis of Feudalism in Mongolia." Translated by Owen Lattimore. *Modern Asian Studies* 1, no. 3 (1967): 265–281.

Ojima, Dennis and Togtohyn Chuluun. "Policy Changes in Mongolia: Implications for Land Use and Landscapes." In *Fragmentation in Semi-Arid and Arid Landscapes: Consequences for Human and Natural Systems*. Edited by K. A. Galvin, R. S. Reid, R. H. Behnke, Jr., and N. T. Hobbs, 179–194. Dordrecht, The Netherlands: Springer, 2010.

Pürev, O. "Khalkhyn Setsenkhan aimgiin borzhigin setsen vangiin khoshuuny tüükhen gazarzüi," *Borzhigin sudlal* 1 (Ulaanbaatar, 1999), as cited in L. Altanzaia, "Khalkhyn Aimag Khoshuudyn nutgiin zakhyn tukhai asuudalkh," *Tüükhiin Setgüül* (Ulaanbaatar, 2003): 84–92.

Rachewiltz, Igor de. "Yeh-lü Ch'u-ts'ai (1189–1243), Yeh-lü Chu (1221–1285)." In *In the Service of the Khan: Eminent Personalities of the Early Mongol-Yüan Period (1200–1300)*. Edited by Igor de Rachewiltz, Hok-lam Chan, Hsiao Ch'i-ch'ing, and Peter W. Geier, 136–175. Wiesbaden: Harrassowitz, 1993.

Reading, Richard P., Don Bedunah, and Sukh Amgalanbaatar. "Conserving Mongolia's Grasslands, with Challenges, Opportunities, and Lessons for North America's Great Plains." *Great Plains Research* 20 (Spring 2010): 85–108.

Reeves, Jeffrey. "Mongolia's Environmental Security: Chinese Unconscious Power and Ulaanbaatar's State Weakness." *Asian Survey* 51, no. 3 (May/Jun 2011): 453–471.

Robinson, Sarah, Peter Finke, and Bettina Hamann. "The Impacts of De-Collectivization on Kazak Pastoralists. Case Studies from Kazakstan, Mongolia and the People's Republic of China." *Journal of Central Asian Studies* 4, no. 2 (Spring/Summer 2000): 2–33.

Róna-Tas, A. "Some Data on the Agriculture of the Mongols." In *Opuscula Ethnologica Memoriae Ludovici Biro Sacra*. Edited by Tibor Bodrogi and L. Boglár, 443–472. *Opuscula* Budapest: Akadémiai Kiadó, 1959.

Rowe, William T. "Owen Lattimore, Asia, and Comparative History." *The Journal of Asian Studies* 66, no. 3 (Aug 2007): 759–786.

Russell, Andrew and Gillian Wallace. "Irresponsible Ecotourism." *Anthropology Today* 20, no. 3 (Jun 2004): 1–2.

Serruys, Henry. "A Question of Land and Landmarks between the Banners Otoγ and Üüsin (Ordos), 1904–1906." *Zentralasiatische Studien* 13 (1979): 215–237.

———. "Altan-qaγan." In *Dictionary of Ming Biography 1368–1644*, vol. 1. Edited by L. Carrington Goodrich and Chaoying Fang, 6–9. New York: Columbia University Press, 1976.

———. "Landgrants to the Mongols in China: 1400–1460." In *The Mongols and Ming China: Customs and History*, 394–405. London: Variorum Reprints, 1987. Originally published in *Monumenta Serica* XXV, 1966.

———. "Notes on a Chinese Inscription of 1606 in a Lamaist Temple in Mai-ta-chao, Suiyüan." *Journal of the American Oriental Society* 78, no. 2 (Apr/Jun 1958): 101–113.

Shinjilt. "Pasture Fights, Mediation, and Ethnic Narrations: Aspects of the Ethnic Relationship between the Mongols and the Tibetans in Qinghai and Gansu." In *The Mongolia-Tibet Interface: Opening New Research Terrains in Inner Asia*. Edited by Uradyn E. Bulag and Hildegard G. M. Diemberger, 337–361. Boston, MA: Brill, 2007.

Shirendyb, B. "Some Aspects of the History of Land Rights in Mongolia." *Journal of the Anglo-Mongolian Society* 3, no. 1 (1976): 44–50 and no. 2 (1976): 41–47.

Sneath, David. "Land Use, the Environment and Development in Post-Socialist Mongolia." *Oxford Development Studies* 31, no. 4 (Dec 2003): 441–459.

———. "Lost in the Post: Technologies of Imagination, and the Soviet Legacy in Post-Socialist Mongolia." *Inner Asia* 5, no. 1 (2003): 39–52.

———. "Proprietary Regimes and Sociotechnical Systems: Rights over Land in Mongolia's 'Age of the Market.'" In *Property in Question: Value Transformation in the Global Economy*. Edited by Katherine Verdery and Caroline Humphrey, 161–182. New York: Berg, 2004.

———. "The Rural and the Urban in Pastoral Mongolia." In *Mongols from Country to City: Floating Boundaries, Pastoralism and City Life in the Mongol Lands*. Edited by Ole Bruun and Li Narangoa, 140–161. Copenhagen: Nordic Institute of Asian Studies, 2006.

Sternberg, Troy. "Unravelling Mongolia's Extreme Winter Disaster of 2010." *Nomadic Peoples* 14, no. 1 (2010): 72–86.

Suttie, J. M. "Case Study 8. Hay from Natural Pasture in Mongolia (The Change from Cooperative to Private Stock-Rearing in a Purely Pastoral Economy)." In *Hay and Straw Conservation for Small-Scale Farming and Pastoral Conditions*, 237–243. Rome: Food and Agriculture Organization of the United Nations, 2000.

———. "Grazing Management in Mongolia." In *Grasslands of the World*. Edited by J. M. Suttie, S. G. Reynolds, and C. Batello, 265–304. Rome: Food and Agriculture Organization of the United Nations, 2005.

Swift, Jeremy, "Rural Development: The Livestock Sector." In *Poverty and Transition to a Market Economy in Mongolia*. Edited by Keith Griffin, 104–133. New York: St. Martin's Press, 1995.

"The Last Best Place." *The Economist* (Dec 21, 2002): 58–60.

Tsogtbaatar, J. "Deforestation and Reforestation Needs in Mongolia." *Forest Ecology and Management* 201 (2004): 57–63.

Tumurjav, Myatavyn. "Traditional Animal Husbandry Techniques Practiced by Mongolian Nomadic People." In *Mongolia Today: Science, Culture, Environment and Development*. Edited by Dendeviin Badarch, Raymond A. Zilinskas, and Peter J. Balint, 86–113. New York: RoutledgeCurzon, 2003.

Wheeler, Alan. "Moralities of the Mongolian 'Market': A Genealogy of Trade Relations and the *Zah Zeel*." *Inner Asia* 6, no. 2 (2004): 215–238.

Wietze, Lise, Sebastiaan Hess, and Byamba Purev. "Pastureland Degradation and Poverty among Herders in Mongolia: Data Analysis and Game Estimation." *Ecological Economics* 58, no. 2 (2006): 350–364.

Yeh, Emily T. "Tibetan Range Wars: Spatial Politics and Authority on the Grasslands of Amdo." *Development and Change* 34, no. 3 (2003): 499–523.

Young, Oran R. "Land Use, Environmental Change, and Sustainable Development: The Role of Institutional Diagnostics." *International Journal of the Commons* 5, no. 1 (Feb 2011): 66–85.

Zhang, MunkhDalai A., Elles Borjigin, and Huiping Zhang. "Mongolian Nomadic Culture and Ecological Culture: On the Ecological Reconstruction in the Agro-Pastoral Zone in Northern China." *Ecological Economics* 62, no. 1 (Apr 2007): 19–26.

Online and Print Periodicals, Websites, and Reports

Note that because Mongolian news websites are frequently cited throughout the book, separate news articles are not listed by author and title below. The Mongolian news websites are, however, listed.

Almaganbetov, Nurlan. "The Impact of Land Reform on Economic Development in Kazakhstan." Central European University Center for Policy Studies: Open Society Institute (2004–2005), www.policy.hu/document/200808/almaganbetov.pdf

BirdLife Asia. "Safeguarding Important Areas of Natural Habitat in Mongolia alongside Economic Development." Washington, DC: World Bank, Mongolia Discussion Papers, East Asia and Pacific Region Sustainable Development Department (Jan 2009), http://siteresources.worldbank.org/INTMONGOLIA/Resources/1_Safeguarding_important_areas_ENG.pdf.

Crisp, Nicholas, John Dick, and Michael Mullins. *Mongolia: Forestry Sector Review.* Victoria, BC: The World Bank (Apr 2004), http://siteresources.worldbank.org/INTMONGOLIA/Resources/eng_version.pdf.

"Death Sentence over Inner Mongolia Herder Killing." *BBC News*, http://www.bbc.co.uk/news/world-asia-pacific-13697612.

Dietz, A. J., E. Amgalan, T. Erdenechuluun, and S. Hess. "Carrying Capacity Dynamics, Livestock Commercialisation and Land Degradation in Mongolia's Free Market Era." Amsterdam: Institute for Environmental Studies, PREM (Poverty Reduction and Environmental Management) Working Paper (Jun 2, 2005), http://www.prem-online.org/archive/2/doc/PREM%20WP%2005–10.pdf.

"Draft Law of Mongolia on Pasture Land," www.greengold.mn/media/... 091103%20Pasture%20Law%20Law%Draft_eng.pdf.

Farmers Knowledge Association (NGO) for the Ministry of Food and Agriculture for the Foreign Investors' Forum in Mongolia. "Information on Mongolian Agriculture" (Jun 2004), www.owc.org.mn/fermerdem/english/monagro/crop_2004.html.

Foundation for the Preservation and Protection of the Przewalski Horse website, http://www.treemail.nl/takh/index.htm.

French, Howard. "The Next Empire?" *The Atlantic* (May 2010), http://www.theatlantic.com/magazine/archive/2010/05/the-next-empire/8018/.

GEF Small Grants Programme website, http://sgp.undp.org/web/projects.

Gün-Galuut Nature Reserve website, http://www.argalipark.com.

IFAD Rural Poverty Reduction Programme in Mongolia. "2005 Progress Report," www.rprmongolia.mn/pdf/2005_annual_report_en.pdf.

Ikh Nart Nature Reserve website, http://www.ikhnart.com/ecotourism.html.

Impacts of Well Rehabilitation and Human Intrusion on Khulan (Wild Ass) and Other Threatened Species in the Gobi Desert, http://khulan.autonomyproductions.com/files/IntroMethods.pdf.

"Law of Mongolia on Land," http://www.investmongolia.com/law28.pdf.

"Law of Mongolia on Land/Revised Version/June 7, 2002," www.nvcmongolia.com;nvce/download/Land.pdf.

Lim, Louisa. "Mongolians Seek Fortune in Gold, But at a Cost." *NPR online* (Sep 7, 2009), http://www.npr.org.

Lovgren, Stefan. "Mongolia Gold Rush Destroying Rivers, Nomadic Lives." *National Geographic News* (Oct 17, 2008), http://news.nationalgeographic.com/news/pf/10277748.html.

Majer, Zsuzsa. "Continuation or Disjuncture with the Past and the Tibetan Buddhist Tradition." *The Silk Road* 7 (2009), www.silk-road.com/toc/newsletter.html.

Mercy Corps Mongolia website, http://www.mercycorps.org.mn.

Millennium Challenge Account Mongolia website, http://www.mca.mn/?=eng/Project/PeriUrban.

Mongolian Ministry of Food and Agriculture website, "Food and Agriculture Policy of Government of Mongolia," http://gate1.pmis.gov.mn/mofa/english/idea_rural.htm.

Mongol Messenger (weekly newspaper printed in Ulaanbaatar).

Mongolia (monthly magazine published in Ulaanbaatar in the socialist period).

Mongoliathisweek, http://www.mongoliathisweek.mn/lanl.htm.

Mongolia Web News, http://www.mongolia-web.com.

"Mongolian Nomadism to 'Die Out,'" *BBC News* (Oct 24, 2003), http://newsvote.bbc.co.uk.

Mongolian Society for Range Management, *Livelihood Study of Herders in Mongolia*. Swiss Agency for Development and Cooperation website (Jun 2010), http://www.swiss-cooperation.admin.ch/...resource_en_184798.pdf.

MSN MoneyCentral, http://news.moneycentral.msn.com.

"No More Nomads in Mongolia, Prime Minister Says." *Far Eastern Economic Review* (May 23, 2001), http://www.feer.com/breaking_news/010523a.html.

Nomadic Expeditions website, http://www.nomadicexpeditions.com.

"Nomads No More: A Steppe-Land Struggles with New Riches." *The Economist* (Oct 21, 2010), http://www.economist.com/node/17312256?story_id=17312256.

Official Tourism Website of Mongolia. "Tourism Director Wants More Well-Informed Guides," http://www.mongoliatourism.gov.mn.

Olsen, Sandra L. "Nomadic Pastoralists, Their Livestock, and Their Landscape," www.carnegiemnh.org/anthro/olsen_mongolia1.html.

Open Society Forum website, www.opensocietyforum.mn.

Pack for a Purpose website, http://www.packforapurpose.org.

River Movements website, http://www.rivermovements.org.

Sant Maral Foundation website, http://www.santmaral.mn/en.
Southern Mongolian Human Rights Information Center. "Herders Take to the Streets, Four Arrested," www.smhric.org/news.
"Special Report: Mongolia's Fabled Mine Stirs Asian Frontier." Reuters (Oct 12, 2010), http://www.alternet.org/thenews/newsdesk/SGE685S060.htm
Steppe Nomads Tourist Camp website, http://www.mongoliagercamp.com.
Swift, Jeremy J. "Case Study: Institutionalizing Pastoral Risk Management in Mongolia: Lessons Learned" (2007), http://www.fao.org/NR/clim/docs/clim_070601_en.pdf
Swiss Agency for Development and Cooperation in Mongolia website, www.sdc.mn.
Teleki, Krisztina. "Building on Ruins, Memories and Persistence: Revival and Survival of Buddhism in the Mongolian Countryside." *The Silk Road* 7 (2009), www.silk-road.com/toc/newsletter.html.
Tibet News Digest, www.tibetinfonet.net.
UB Post (online version), http://ubpost.mongolnews.mn.
United Nations Development Programme. "Human Development Report: Mongolia 2003," http://hdr.undp.org/en/reports/national/asiathepacific/mongolia/mongolia_2003_en.pdf.
United Nations Development Programme. "Mongolia Human Development Report 2011: From Vulnerability to Sustainability: Environment and Human Development," http://www.undp.mn/nhdr2011/NHDR_report_english_2011.pdf.
United Nations Development Programme. *UNDP Mongolia: News: Sustainable Natural Resource Management*, http://www.undp.mn/news-snrm-200510.html.
US Department of State, "Background Note: Mongolia," http://www.state.gov/r/pa/ei/bgn/2770.htm.
USAID Mongolia website, http://mongolia.usaid.gov.
Wallace, Vesna. "Mediating the Power of Dharma: The Mongols' Approaches to Reviving Buddhism in Mongolia." *The Silk Road* 6, no. 1 (Summer 2008), www.silk-road.com/toc/newsletter.html.
Watts, Jonathan. "Herder's Death Deepens Tensions in Inner Mongolia." *The Guardian* (May 27, 2011), http://www.guardian.co.uk/world/2011/may/27/tensions-miners-inner-mongolia.
World Bank. "Index-Based Livestock Insurance Project," http://www-wds.worldbank.org.
World Bank. "Mongolia: Environment Monitor 2004," http://www-wds.worldbank.org.
World Bank. *Mongolia Quarterly* (Jan 28, 2008), http://siteresources.worldbank.org/INTMONGOLIA/Resources/MNQtrlyFeb2008.pdf.

Index

Page numbers in *italics* refer to figures and illustrations.

9781137269652